Mandy Cook

The neurodegenerative Drosophila melanogaster AMPK mutant loechrig

Mandy Cook

The neurodegenerative Drosophila melanogaster AMPK mutant loechrig

The fly mutant that helps uncover the relationship between AMPK, actin dynamics and neuronal cell death

Südwestdeutscher Verlag für Hochschulschriften

Impressum / Imprint
Bibliografische Information der Deutschen Nationalbibliothek: Die Deutsche Nationalbibliothek verzeichnet diese Publikation in der Deutschen Nationalbibliografie; detaillierte bibliografische Daten sind im Internet über http://dnb.d-nb.de abrufbar.
Alle in diesem Buch genannten Marken und Produktnamen unterliegen warenzeichen-, marken- oder patentrechtlichem Schutz bzw. sind Warenzeichen oder eingetragene Warenzeichen der jeweiligen Inhaber. Die Wiedergabe von Marken, Produktnamen, Gebrauchsnamen, Handelsnamen, Warenbezeichnungen u.s.w. in diesem Werk berechtigt auch ohne besondere Kennzeichnung nicht zu der Annahme, dass solche Namen im Sinne der Warenzeichen- und Markenschutzgesetzgebung als frei zu betrachten wären und daher von jedermann benutzt werden dürften.

Bibliographic information published by the Deutsche Nationalbibliothek: The Deutsche Nationalbibliothek lists this publication in the Deutsche Nationalbibliografie; detailed bibliographic data are available in the Internet at http://dnb.d-nb.de.
Any brand names and product names mentioned in this book are subject to trademark, brand or patent protection and are trademarks or registered trademarks of their respective holders. The use of brand names, product names, common names, trade names, product descriptions etc. even without a particular marking in this works is in no way to be construed to mean that such names may be regarded as unrestricted in respect of trademark and brand protection legislation and could thus be used by anyone.

Coverbild / Cover image: www.ingimage.com

Verlag / Publisher:
Südwestdeutscher Verlag für Hochschulschriften
ist ein Imprint der / is a trademark of
AV Akademikerverlag GmbH & Co. KG
Heinrich-Böcking-Str. 6-8, 66121 Saarbrücken, Deutschland / Germany
Email: info@svh-verlag.de

Herstellung: siehe letzte Seite /
Printed at: see last page
ISBN: 978-3-8381-3489-5

Zugl. / Approved by: Würzburg, U, Diss., 2012

Copyright © 2012 AV Akademikerverlag GmbH & Co. KG
Alle Rechte vorbehalten. / All rights reserved. Saarbrücken 2012

TABLE OF CONTENTS

Table of Figures iii
List of Tables iv
Summary 1
Zusammenfassung 3

1 Introduction 5
1.1 Why use *Drosophila melanogaster* as model organism 6
1.2 The *Drosophila* mutant *loechrig* 8
1.3 AMPK - and why it is so important 9
   1.3.1 AMPK upstream kinases 13
   1.3.2 AMPK mutation causing disease in humans 13
1.4 The LOE Protein 16
1.5 Protein Prenylation 18
   1.5.1 The Isroprenoid Pathway 19
1.6 The RHO Pathway and the Role of RHO 21
   1.6.1 Actin 23
   1.6.2 LIM-Kinase and Cofilin 25

2 Material Methods 26
2.1 Material 26
   2.1.1 Fly Strains 26
   2.1.2 Antibodies 28
      2.1.2.1 Primary 28
      2.1.2.2 Secondary 29
   2.1.3 Ladders and Enzymes 29
   2.1.4 Kits 29
2.2 Methods 30
   2.2.1 Working with Flies 30
      2.2.1.1 Fly keeping 30
      2.2.1.2 Paraffin sections 30
      2.2.1.3 Fast phototaxis assay 31
   2.2.2 Working with Proteins 32
      2.2.2.1 Western Blot 32
      2.2.2.2 Visualizing F-actin and G-actin levels 33
   2.2.3 Immunoprecipitation and subcellular fractionation 34
   2.2.4 Working with primary cells 34
      2.2.4.1 Measuring neurite length and movement of mitochondria 35

3 Results 36
3.1 Interactions with RHO and ROK 37

    3.1.1   loe shows more isoprenylation of Rho1 37
    3.1.2   More Rho is located at the membrane in *loe* 38
    3.1.3   The influence of ROK1 on the neurodegenerative phenotype in *loe* 41
3.2   **Effect of the Proteins LIMK and Slingshot on *loe*** 42
3.3   **The Protein Cofilin** 46
    3.3.1   The influence of cofilin on the neurodegenerative phenotype in *loe* 46
    3.3.2   Up-regulation of inactive cofilin in *loe* 48
3.4   **The Role of Actin** 50
3.5   ***loe* neurons show a change in neurite outgrowth and organelle transport** 52
    3.5.1   Outgrowth of neurites is affected in *loe* 52
    3.5.2   Slower transport through neurites in loe 55
3.6   **Behavioral phenotype in *loe*** 57
3.7   **AMPK γ RNAi shows a degenerative phenotype with various drivers** 63
    3.7.1   Phenotype in UAS-AMPK γ RNAi driven with GMR II – Gal4 and GMR III – Gal4 and Ey-Gal4   63
    3.7.2   Phenotype in UAS AMPK γ RNAi driven neuronal drivers 65
3.8   **Cholesterol Influences neurodegeneration** 66

4 Discussion 69
4.1   ***loe* interferes with the Rho pathway** 70
4.2   ***loe* has more inactive cofilin** 74
4.3   **Cholesterol effects the neurodegeneration phenotype** 75
4.4   **Neurite outgrowth and transport is affected in *loe*** 76

5 References 80

6 Abbreviations 85

List of Publications 87

TABLE OF FIGURES

Figure 1 - The common fruit fly- *Drosophila melanogaster* 7
Figure 2 - Neurodegenerative *loe* Phenotype 9
Figure 3 - The Role of AMPK in the cell 11
Figure 4 - Myocardium section of patient with AMPK mutation 14
Figure 5 - Heart schematic of WPW disease 15
Figure 6 - P-element insertion in the *loe* gene 16
Figure 7 – Homology of AMPK γ-subunits of different species 17
Figure 8 – Isoprenoid pathway 20
Figure 9 – Rho Pathway 22
Figure 10 – Actin tread milling in the growth cone 23
Figure 11 – *loe* affects the levels of farnesylation of Rho1 37
Figure 12 – More Rho is found at the membrane in *loe* 39
Figure 13 – A Rho- Kinase mutant decreases neurodegeneration 42
Figure 14 – Schematic of the GAL4- UAS system in *Drosophila melanogaster* 43
Figure 15 – Increased activation of LIMK or SSH results in enhanced neurodegeneration in *loe* 44
Figure 16 – Effects of TSR on *loe* neurodegeneration 47
Figure 17 – Increased level of P-Cofilin and SSH in *loe* 49
Figure 18 – Loe shows increased level of fibrilar actin 51
Figure 19 – *loe* neuron cultured for twenty-four hours 53
Figure 20 – Neurite length of larval brain neurons 54
Figure 21 – Velocity of mitochondria and number of processes in neurons 56
Figure 22 – Phototaxis assay with *loe* flies 57
Figure 23 – Phototaxis assay with *loe tsr* and *loe rho* flies 58
Figure 24 – Phototaxis assay with TSR overexpression 59
Figure 25 – Phototaxis assay with SSH overexpression 60
Figure 26 – Life span in *loe* and *rho loe* flies 61
Figure 27 – Cell death in the eye causes by AMPK RNAi 64
Figure 28 – Degeneration in neurons caused by the AMPK RNAi line 65
Figure 29 – Cholesterol ester concentration in *loe* 67
Figure 30 – Feeding assay with cholesterol free food 68

LIST OF TABLES

Table 1- List of flies 26
Table 2- List of primary antibodies 28
Table 3- List of secondary antibodies 29
Table 4- List of ladders 29
Table 5- List of kits 29

SUMMARY

In this thesis the *Drosophila* mutant *loechrig* (*loe*), that shows progressive degeneration of the nervous system, is further described. *Loe* is missing a neuronal isoform of the protein kinase AMPK γ subunit (AMP-activated protein kinase- also known as SNF4Aγ) The heterotrimeric AMPK controls the energy level of the cell, which requires constant monitoring of the ATP/AMP levels. It is activated by low energy levels and metabolic insults like oxygen starvation and regulates multiple important signal pathways that control cell metabolism. Still, its role in neuronal survival is unclear.

One of AMPK's downstream targets is HMGR (hydroxymethylglutaryl-CoA-reductase), a key enzyme in cholesterol and isoprenoid synthesis. It has been shown that manipulating the levels of HMGR affects the severity of the neurodegenerative phenotype in *loe*. Whereas the regulatory role of AMPK on HMGR is conserved in *Drosophila*, insects cannot synthesize cholesterol *de novo*. However, the synthesis of isoprenoids is a pathway that is evolutionarily conserved between vertebrates and insects.

Isoprenylation of target proteins like small G-proteins provides a hydrophobic anchor that allows the association of these proteins with membranes and following activation. This thesis shows that the *loe* mutation interferes with the prenylation of Rho1 and the regulation of the LIM kinase pathway, which plays an important role in actin turnover and axonal outgrowth.

The results suggest that the mutation in LOE, causes hyperactivity of the isoprenoid synthesis pathway, which leads to increased farnesylation of RHO1 and therefore higher levels of phospho-cofilin. A mutation in Rho1 improves the neurodegenerative phenotype and life span. The increased inactive cofilin amount in *loe* leads to an up regulation of filamentous actin. Actin is involved in neuronal outgrowth and experiments analyzing *loe* neurons gave valuable insights into a possible role of AMPK and accordingly actin on neurite growth and stability.

It was demonstrated that neurons derived from *loe* mutants exhibit reduces axonal transport suggesting that changes in the cytoskeletal network caused by the effect of *loe* on the Rho1 pathway lead to disruptions in axonal transport and subsequent neuronal death. It also shows that actin is not only involved in neuronal outgrowth, its also important in maintenance of neurons, suggesting that interference with actin dynamics leads to progressive degeneration of neurons. Together, these results further support the importance of AMPK in neuronal function and survival and provide a novel functional mechanisms how alterations in AMPK can cause neuronal degeneration.

ZUSAMMENFASSUNG

In dieser Doktorarbeit wird die *Drosophila* Mutante *loechrig* (*loe*), die progressive Degeneration des Nervensystems aufweist, weiter beschrieben. In der *loe* Mutante fehlt eine neuronale Isoform der γ- Untereinheit der Proteinkinase AMPK (AMP-activated protein kinase). Die heterotrimere AMPK (auch als SNF4Aγ bekannt) kontrolliert das Energieniveau der Zelle, was ständiges Beobachten des ATP/AMP- Verhältnis erfordert. AMPK wird durch niedrige Energiekonzentrationen und Beeinträchtigungen im Metabolismus, wie zum Beispiel Sauerstoffmangel, aktiviert und reguliert mehrere wichtige Signaltransduktionswege, die den Zellmetabolismus kontrollieren. Jedoch ist die Rolle von AMPK im neuronalen Überleben noch unklar.

Eines der Proteine, dass von AMPK reguliert wird, ist HMGR (hydroxymethylglutaryl-CoA- reductase), ein Schlüsselenzym in der Cholesterin- und Isoprenoidsynthese. Es wurde gezeigt, dass wenn die Konzentration von HMGR manipuliert wird, auch der Schweregrad des neurodegenerativen Phänotyps in *loe* beeinflusst wird. Obwohl die regulatorische Rolle von AMPK auf HMGR in *Drosophila* konserviert ist, können Insekten Cholesterin nicht *de novo* synthetisieren. Dennoch ist der Syntheseweg von Isoprenoiden zwischen Vertebraten und Insekten evolutionär konserviert.

Isoprenylierung von Proteinen, wie zum Beispiel von kleinen G-Proteinen, stellt den Proteinen einen hydophobischen Anker bereit, mit denen sie sich an die Zellmembran binden können, was in anschließender Aktivierung resultieren kann. In dieser Doktorarbeit wird gezeigt, dass die *loe* Mutation die Prenylierung von Rho1 und den LIM-Kinasesignalweg beeinflusst, was eine wichtige Rolle im Umsatz von Aktin und axonalem Auswachsen spielt.

Die Ergebnisse weisen darauf hin, dass die Mutation in LOE, Hyperaktivität des Isoprenoidsynthesewegs verursacht, was zur erhöhten Farnesylierung von Rho1 und einer dementsprechend höheren Konzentration von Phospho- Cofilin führt.

Eine Mutation in Rho1 verbessert den neurodegenerativen Phänotyp und die Lebenserwartung von *loe*. Der Anstieg vom inaktiven Cofilin in *loe* führt zu einer Zunahme von filamentösen Aktin. Aktin ist am Auswachen von Neuronen beteiligt und Experimente in denen *loe* Neurone analysiert wurden, gaben wertvolle Einblicke in eine mögliche Rolle die AMPK, und dementsprechend Aktin, im Neuronenwachstum spielt.

Des Weiteren wurde demonstriert, dass Neurone, die von der *loe* Mutante stamen, einen verlangsamten axonalen Transport aufweisen, was darauf hinweist dass Veränderungen, die durch den Einfluss von *loe* auf den Rho1 Signalweg im Zytoskelettnetzwerk hervorgerufen wurden, zur Störung des axonalen Transports und anschließenden neuronalen Tod führen. Es zeigte außerdem, dass Aktin nicht nur am neuronalen Auswachsen beteiligt ist, sondern
auch wichtig für die Aufrechterhaltung von Neuronen ist. Das bedeutet, dass Änderungen der Aktindynamik zur progressiven Degeneration von Neuronen führen kann.

Zusammenfassend unterstreichen diese Ergebnisse die wichtige Bedeutung von AMPK in den Funktionen und im Überleben von Neuronen und eröffnen einen neuartigen funktionellen Mechanismus in dem Änderungen in AMPK neuronale Degeneration hervorrufen kann.

1 INTRODUCTION

Neurodegenerative diseases are becoming increasingly present in our modern society due to the demographic changes and longer life expectancy. Neurodegenerative diseases are caused by progressive loss of specific neurons and are typically age-related human disorders with important pathological and clinical similarity.

For instance, Alzheimer's disease, which is the most common form of dementia and the 6th leading cause of death (Alzheimer's Association; www.alz.org), affects 35 million people worldwide (Querfurth and LaFerla, 2010) and with the increase in life span this number is predicted to quadruple by 2050 (Brookmeyer *et al.*, 2007).

Other well-characterized neurodegenerative diseases include Parkinson's disease (PD) and Huntington's disease. PD is the most frequent movement disorder known and is caused by a progressive loss of dopaminergic neurons and the consequent depletion of the neurotransmitter dopamine in the striatum (connected to the basal ganglia). There have been several genes linked to the familial form of PD, however the majority of the cases are sporadic and the cause is unknown (Correia *et al.*, 2012). In Huntington's disease, which is also an age related disorder that affects brain cells, an extended CAG stretch mutation in the *huntingtin* gene leads to a Poly-Q repetition that causes a loss of neurons in the striatum and cortex (Bates, 2005).

The mechanisms that lead to cell death in most neurodegenerative diseases, including the three mentioned above, are so far irreversible. Therefore it is so important to understand which pathways are involved and how they are affected in order to improve existing treatments and finding a prevention to help people that suffer from these diseases.

In many neurodegenerative diseases, the identification of mutations associated with familial cases, like in Huntington's disease, has allowed investigators to develop *in vitro* and *in vivo* model systems to define the cellular and molecular aberrations associated with the mutant gene product. Understanding the molecular processes that can cause neurodegeneration are of prime importance and will be addressed in this thesis with the aid of a neurodegenerative fly model.

1.1 Why use *Drosophila melanogaster* as model organism

Drosophila melanogaster, also known as the common fruit fly (Figure 1) is an excellent model organism for the study of genes and molecular pathways that are known to be involved in the development of human diseases. Surprisingly, a lot of genes are conserved between mammals and diptera and approximately 75% of human disease-causing genes are believed to have a functional homolog in the fly (Pandey and Nichols, 2011). In fact, the gene *loe,* which will be discussed in this work is conserved in humans and it has been closely related to several rare diseases in humans (look at 1.3.2).

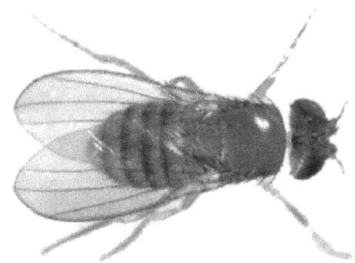

Figure 1 - The common fruit fly- *Drosophila melanogaster*

Given that the fly has a short reproduction cycle (10 days), it is fairly easy to generate several generations in a relatively short time. The fruit fly also has, compared to humans, a small genome (flies-14.000 genes, humans 20.000-30.000 genes) and a high number of genetic tools with numerous markers that allow us to generate various mutants, which can then be studied. The fly can therefore help to find new insights into the basic mechanisms and genetic pathways important for various research fields.

When pathways in the development of the nervous system or organogenesis are disrupted similar defects in either vertebrates or invertebrates can be observed. Human diseases like Huntington's disease (Fortini and Bonini, 2000), hereditary spastic paraplegia, which is an NTE-related motor neuron disorder (Bettencourt da Cruz *et al.*, 2008) and Alzheimer's disease (Crowther *et al.*, 2005) have been analyzed effectively using *Drosophila* as a model genetic system.

Another benefit of working with flies is the big selection of genetic tools that are available for the *Drosophila* model. For example, the UAS-GAL4-System, which will be described in the third chapter (Figure 14) is used daily in fly laboratories. It is a very powerful tool for the geneticist, because it allows to express, overexpress or down regulate a gene of interest in almost every tissue or body part of the fly.

1.2 The *Drosophila* mutant *loechrig*

The *loe* mutant was generated in a P-element (transposon in *Drosophila*) insertion collection (Deák *et al.*, 1997) of genes on the third chromosome and subsequently characterized in the Kretzschmar laboratory.

Particularly noticeable about the new mutant was the neurodegenerative phenotype in the brain, which is caused by necrotic cell death of neurons. The brains of newly eclosed *loe* flies show no apparent defects, but within the next 4 days of adult life they display spongiform lesions that keep enlarging with time. The "big holes" phenotype (Figure 2B) is the reason for naming the mutant *löchrig*, which means "full of holes" in German. The increasing number of dying neurons in the brain leads to a premature death of the flies after approximately three weeks.

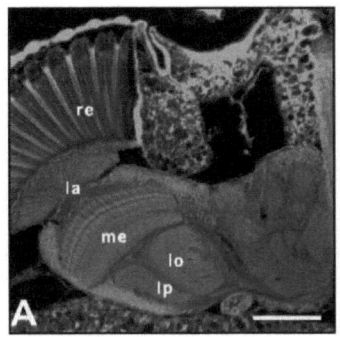 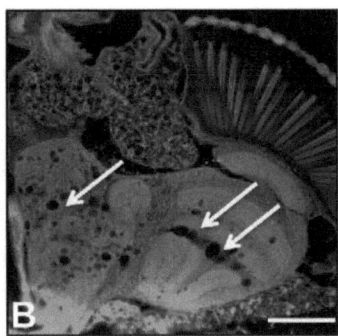

Figure 2 - Neurodegenerative *loe* Phenotype
Paraffin section of an adult fly head (5d old): wild- type (re- retina, la- lamina, me- medulla, lo- lobula, lp- lobula plate) (A) and vacuolization (white arrows) in *loe* (B). Scale bar= 50μm

1.3 AMPK - and why it is so important

Adenosine triphosphate (ATP), which cleaves to adenosine diphosphate (ADP) thereby releasing energy, is critical for maintaining energy homeostasis within the cell. Every cellular reaction contributes to energy consumption. One example of this is nucleosome remodeling, which leads to DNA unwrapping (Blossey and Schiessel, 2011) or release, e.g., oxidation of glucose. It is essential for the cell to have the energy flow constantly monitored by sensing the ATP level. Because this is a key function of AMPK (AMP- activated protein kinase) it is also called "the energy sensor of the cell" (Figure 3).

The heterotrimeric protein kinase AMPK is activated by an increased ratio of AMP to ATP, which results in the activation of energy-producing pathways (catabolic reactions), like glycolysis, and the suppression of energy-consuming ones (anabolic reactions), like cholesterol synthesis, to restore ATP levels. By regulation of AMPK in the hypothalamus of the brain through cytokines, food intake can be controlled (Hardie, 2007), which is critical for drug research in the obesity field. Furthermore, in yeast, AMPK plays an important role in metabolism. It can react to the energy levels with a fast response through direct phosphorylation or a slower response via adaptive transcription. With those responses, AMPK is able to communicate between the metabolic environment and the energy status of the cell. Because of its significant cellular value, AMPK is highly conserved between eukaryotes. Bland et al. has shown that SNF1 (yeast homologue of AMPK) is essential for cellular adaptation to growth on non-glucose carbon sources. (Bland et al., 2010) This emphasizes the close connection between AMPK and glycogen metabolism.

AMP- activated protein kinase consists of one catalytic α and two regulatory β and γ subunits. Mammals have two isoforms for the α- and β- subunit and three isoforms for the γ subunit. That means the AMPK can be assembled to twelve different isoform combinations.
At the tissue level, AMPK trimer composition is extremely varied (Cantó and Auwerx, 2010). The tissue-specific pattern is especially clear for the γ 3 subunit of AMPK. Its expression is mostly restricted to glycolytic skeletal muscle, where it is the most common γ isoform (Mahlapuu et al., 2004).

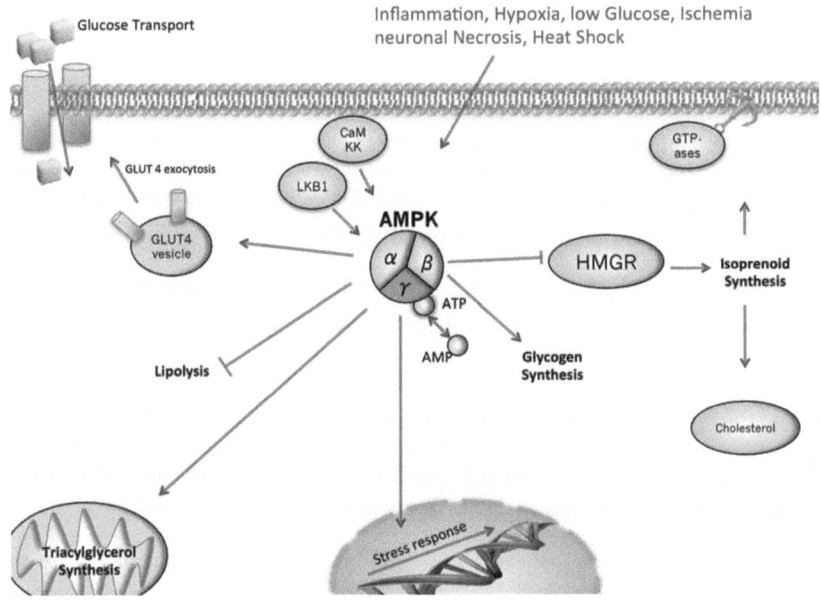

Figure 3 - The Role of AMPK in the cell

The activity of the heterotrimeric protein kinase AMPK is depended on the energy level in the cell. It gets phosphorylated by upstream kinases and activated by a high AMP-ATP ratio. AMPK controls glycogen synthesis and is able to regulate the transcription of a number of gene sets. In response to muscle contraction AMPK can even influence the translocation of GLUT4 and therefore induce glucose uptake. To reduce energy consumption, AMPK switches off anabolic pathways such as triglyceride, fatty acid and cholesterol synthesis and turns on catabolic pathways like glycolysis and lipolysis (fatty acid oxidation).

Each isoform has its own function. The C-terminal region of the α- subunit is required for formation of the complex with the β- and γ- subunit (Crute *et al.*, 1998) and it contains the Thr172 residue, which phosphorylation is required for full enzymatic activity (Hawley *et al.*, 1996). The β subunit has a myristoyl sequence, which leads some researchers to believe that AMPK could undergo a "myristoyl switch" that leads to increased AMPK activity (Steinberg and Kemp, 2009). The β subunit also contains an evolutionary conserved carbohydrate-binding domain (Hudson *et al.*, 2003). This domain enables AMPK to interact with glycogen particles, which is important for its earlier mentioned role in glycogen metabolism. The γ subunit has a very special function- it binds AMP/ATP. The AMP/ATP binding is necessary for AMPK to perform its unique energy sensing function. As long as ATP is bound to the γ subunit (Figure 3), the enzyme activity is low. As soon as AMP replaces ATP, the enzyme activity rises although only slightly at first. With the binding of AMP to the allosteric site in the γ subunit, more phosphorylation of the α subunit takes place, which increases the AMPK activity to more than 1,000-fold (Suter *et al.*, 2006).

1.3.1 AMPK UPSTREAM KINASES

There are only a few known upstream kinases (AMPKK) for AMPK. Calmodulin-dependent kinase kinase (CaMKK), which is highly expressed in the brain, LKB1 and TGF-β-activated kinase-1 (TAK1). The structure and the function of the known AMPKKs is beyond the scope of this thesis, but there is one very interesting fact about LKB1 that should be mentioned. When mutated, LKB1, known as a tumor suppressor, can cause Peutz–Jeghers syndrome (PJS). PJS is a rare genetic disease that is characterized by an increased likelihood of epithelial cancers and intestinal hamartomas. Patients that have inherited this autosomal dominant disease develop benign hamartomatous polyps, especially in the gastrointestinal tract, that will, in most cases develop into malignant tumors (Ji *et al.*, 2007). Another symptom of PJS is cutaneous pigmentation of the mucous membranes (Alessi *et al.*, 2006).

1.3.2 AMPK MUTATION CAUSING DISEASE IN HUMANS

Approximately three-quarters of the candidate human disease genes (929 genes where found in blast analysis against the *Drosophila melanogaster* genome) are clearly related to genes in *Drosophila* (Reiter *et al.*, 2001). Several point mutations in the γ2- subunit isoform are associated with a cardiac glycogen storage phenotype. A rare mutation in PRKAG2, the γ2 isoform of the human AMPK, can lead to Wolff-Parkinson-White syndrome that displays a distinctive cardiac histopathology which leads to arrhythmia and can cause sudden death.

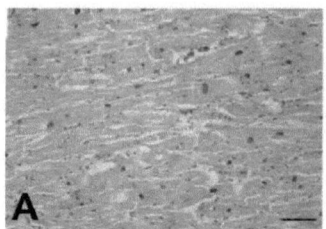 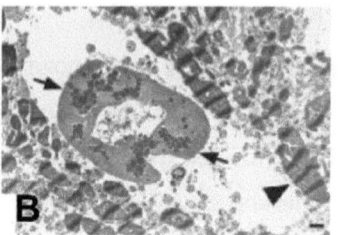

Figure 4 - Myocardium section of patient with AMPK mutation
A: Longitudinal section with vacuolated myoctes of a 26 year old individual who died suddenly (bar = 100μm). B: Electron micrograph shows large, irregular sarcoplasmic inclusion (arrows) within a large vacuole and normal-appearing sarcomeres (arrowhead) (bar = 1μm).(Arad et al., 2002)

Scientist are trying to find out whether a PRKAG2 mutation leads to hypertrophic cardiomyopathy (Burwinkel et al., 2005) or if it is myocardial metabolic storage disease characterized by enlarged myocytes with vacuoles containing glycogen derivatives in which hypertrophy, ventricular pre-excitation and conduction system defects coexist (Arad et al., 2002).

Figure 4 shows a section of muscular tissue with enlarged myocytes and another section with dense packed granular that is characteristic of amylopectin, a nonsoluble product of glycogen metabolism. Glycogen can generate sinus dysfunction and electrophysiological abnormalities if accumulated in conductive tissue.

Introduction

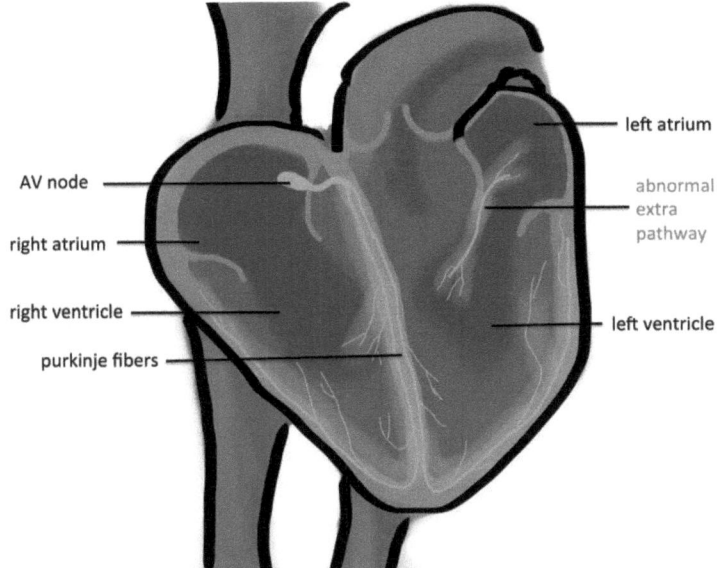

Figure 5 - Heart schematic of WPW disease
An additional conductive pathway between atrium and ventricle leads to preexcitation.

One characteristic of the Wolff-Parkinson-White syndrome is an electrophysiological abnormality, which involves an extra, abnormal electrical pathway between the atrium and the ventricle of the heart that bypasses the atrioventricular (AV) node and disrupts the normal timing of the hearts electrical system (Sidhu *et al.*, 2005). With open-heart surgery, a permanent cure is possible by ablation of the extra pathway.

1.4 The LOE Protein

The fly mutant *loe*, was described in 2002 in the Kretzschmar lab (Tschäpe et al., 2002). LOE, also known as SNF4Aγ, encodes the γ subunit of the fly AMPK. Whereas vertebrates have several genes for each subunit, flies contain only a single gene for each subunit, which simplifies the genetic analysis of AMPK's function in *Drosophila*. However, the γ subunit has at least six alternatively spliced transcripts (Figure 6). The LOE I isoform has a unique N-terminus and is expressed in the brain and needed for the brain's maintenance because it cannot be substituted by other isoforms (Tschäpe et al., 2002).

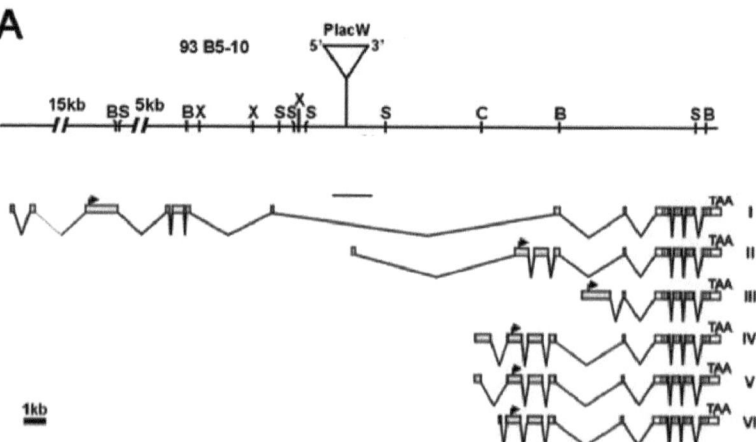

Figure 6 - P-element insertion in the *loe* gene
Shown is the exon/intron structures of the six LOE transcripts (LoeI to LoeVI). The P- element insertion (PlacW) of the *loe* mutant is located in the seventh intron of the Loe I transcript and 38 bp upstream of the transcription start site of LoeII (Tschäpe et al., 2002).

As mentioned before, the AMPK γ subunit has the important role of binding ATP or AMP, which is necessary for assessing the energy status of the cell.

In humans the three different AMPK γ subunit isoforms contain the β binding domain and Bateman domains. The Bateman domain consists of different combinations of CBS (cystathionine-β-synthase) domains, which are AMP/ATP binding domains (Hardie, 2007).

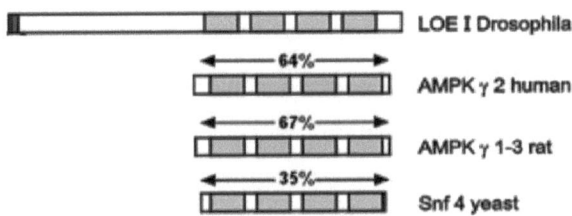

Figure 7 – Homology of AMPK γ-subunits of different species
Homology of the LOE I C-terminus between *Drosophila*, human rat and yeast. Light gray areas indicate the CBS domain, which binds AMP and ATP (Tschäpe et al., 2002).

1.5 Protein Prenylation

In complex organisms, proliferation, differentiation and survival of the cell are regulated by a number of extracellular hormones, growth factors and cytokines. The signal transduction of these factors from the cell surface to the nucleus often requires the presence of small intracellular proteins like small G-proteins. Small G-proteins get switched on when they bind GTP (guanosine triphosphate) and get switched off by binding GDP (guanosine diphosphate). Small G-proteins are linked to the plasma membrane through a hydrophobic anchor. For this group of proteins, biological activity is dependent on localization to the cell membrane. Protein anchoring can be achieved by prenylation of the C-terminus or myristoylation of the N-terminus.

Protein isoprenylation is a post-translational modification in which a downstream product of mevalonate, either farnesyl or geranylgeranyl isoprenoid, is attached to a carboxy-terminal cysteine residue (Goldstein and Brown, 1990). Without prenylation, proteins like Rac, Ras and Rho would be unable to find the right intracellular position, which is a necessary condition for them to change into their active form. To be modified by farnesylation or geranylgeranylation, a sequence of amino acids in a protein must become more hydrophobic and then further modified by methylation. This hydrophobic site allows proteins to bind to the membrane (Glomset *et al.*, 1990).

For example Inglese *et al* demonstrated that the kinase of Rhodopsin (G-protein coupled receptor), which is responsible for light processing in the retina, is unable to anchor the photon-activated rhodopsin when isoprenoid is mutated. Without the isoprenoid anchor, the rhodopsin kinase cannot be translocated to the membrane and consequently not be activated (Inglese *et al.*, 1992).

It has been shown that changes in protein encoding genes that are involved in prenylation can lead to rare diseases like Hutchinson-Gilford progeria syndrome, retinitis pigmentosa and more frequent diseases like cancer (Novelli and D'Apice, 2012). And as early as the 90's it was assumed that manipulation of this regulatory system could be useful in treating certain forms of cancer, as well as heart disease (Goldstein and Brown, 1990). One of the genes, that when mutated can cause cancer is p21, which is the gene product of the protooncogene *ras*. Because p21 requires a farnesyl moiety to bind to the cell membrane, interfering with its function by manipulating its isoprenylation could therefore provide a therapeutic treatment.

1.5.1 THE ISROPRENOID PATHWAY

The isoprenoid pathway, also known as the mevalonate pathway, leads to synthesis of Geranyl-PP and Farnesyl-PP and cholesterol (Figure 8). However, insects do not synthesize *de novo* cholesterol, because they lack at least two enzymes necessary for this synthesis (Gertler *et al.*, 1988). In order to sustain growth and reproduction cholesterol has to be obtained from food, which is then also used to synthesize the molting hormone 20-hydroxyecdysone (20HE) (Clark and Block, 1959).

HMG-CoA-Reductase (HMGR) is a key enzyme in the isoprenoid pathway, it reduces Acetoacetyl- CoA to HMG-CoA. AMPK acts as a negative regulator on HMGR. Therefore it controls the rate of synthesis of downstream products from Acetoacetyl- CoA. Due to AMPK not being fully functional in the *loe* fly, because of the P-element insertion in the γ subunit, the inhibition of HMGR through AMPK should be eliminated. This was confirmed by genetic interactions between *loe* and a *Drosophila* mutant in HMGR (Tschäpe *et al.*, 2002).

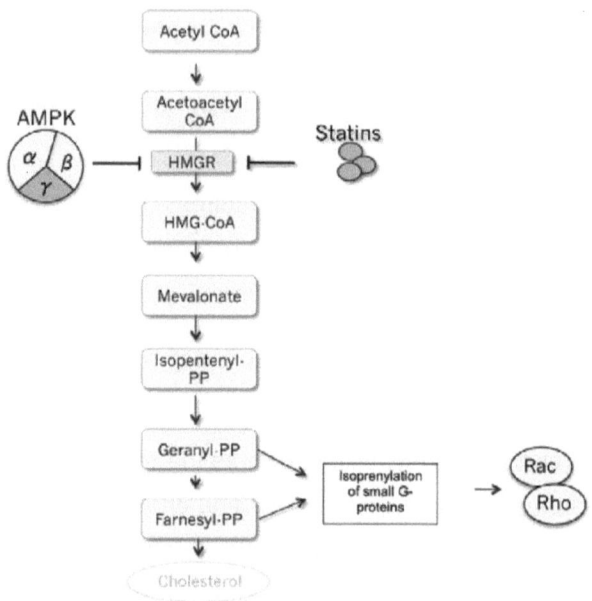

Figure 8 – Isoprenoid pathway
The isoprenoid pathway leads to cholesterol and isoprenoid synthesis. By inhibiting HMGR, AMPK controls the prenylation rate through farnesyltransferase and geranylgeranyltransferase. Target proteins, like RHO will be modified and receive an hydrophopic anchor. (PP-pyrophosphate).

HMGR can be inhibited pharmacologically by a class of drugs called statins (Figure 8), which have been shown to decrease the prevalence of Alzheimer's disease (Wolozin et al., 2000). Tschäpe et al. also tested if statins can suppress the neurodegenerative *loe* phenotype and found indeed an improvement of vacuolization by feeding statins to *loe* flies (Tschäpe et al., 2002). Due to the missing inhibition of HMGR in *loe*, one would expect changes in isoprenylation, which was addressed in this thesis.

1.6 The RHO Pathway and the Role of RHO

Since the prenylation plays an important role in the modification of small G- proteins (e.g. Rho), effects of *loe* on the Rho signaling pathway (Figure 9) were investigated.

Rho GTPases are essential regulators of cytoskeletal reorganization, because they control multiple aspects of growth cone behavior, including growth, branching, turning, retraction and pausing. Growth cones, which are found at the end of neuronal processes, are essential for neuronal outgrowth to connect to their targets and for the formation of neural circuits. In response to environmental guidance signals, they are also required to initiate the neurons extension over long distances.

The Rho pathway is conserved in flies (Ng and Luo, 2004), and Rho GTPase signaling can be modulated by many extra cellular cues that regulate neuronal morphogenesis through interactions with positive regulators

(RhoGEFs), negative regulators (RhoGAPs), Rho GTPases themselves, or downstream effectors (Luo, 2000).

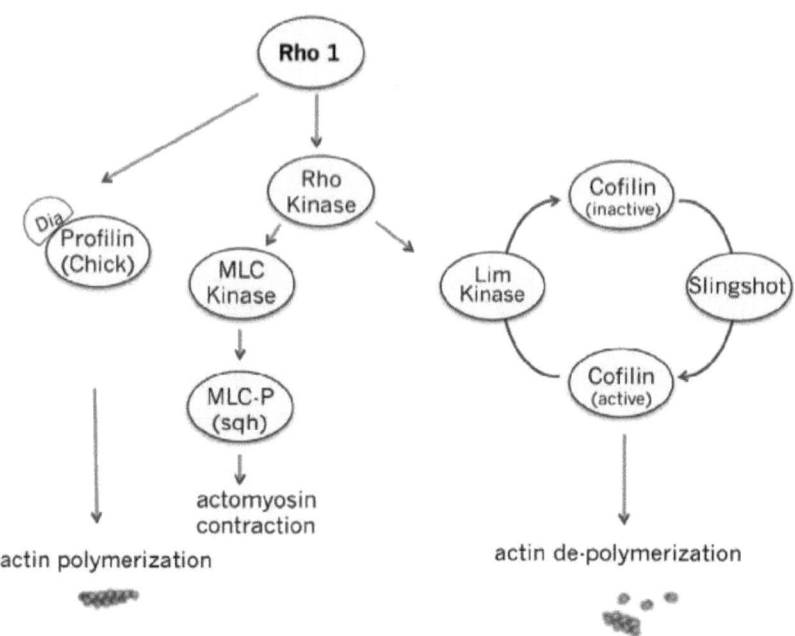

Figure 9 – Rho Pathway
The Rho Pathway controls the actin dynamics by activating/deactivating cofilin and profilin. It also regulates actomyosin contractions. LIM Kinase phosphorylates cofilin, which inhibits it. Slingshot phosphatase activates cofilin by de- phosphorylation. Red circles indicate actin monomers

1.6.1 ACTIN

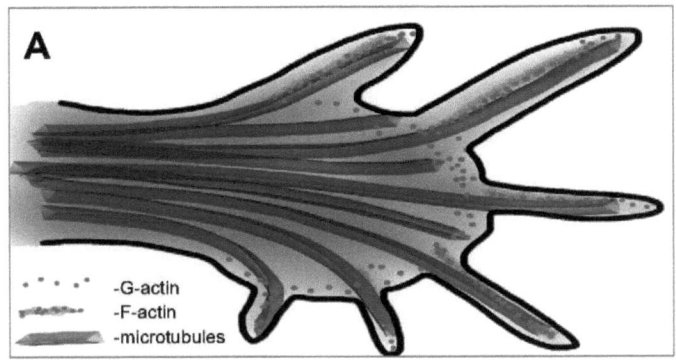

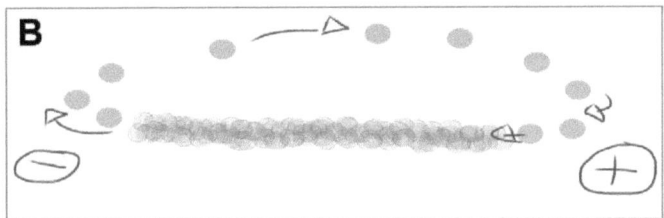

Figure 10 – Actin tread milling in the growth cone

Actin filament plays a major role in the outgrowth of the filopodia of the growth cone (A). Proper functioning of actin recycling (B) is necessary for the filamentous actin to move along. Acting recycling/tread milling requires regulation of actin dephosphorylation on the minus- end and phosphorylation on the plus- end.

Axon elongation and dendritic branching is mediated by the actin and microtubule cytoskeletons. The ability of the cell to change from actin monomers (G-actin) to filaments (F-actin) and vice versa is required to maintain the flexibility that the actin cytoskeleton needs to respond to the outside stimuli. In this process, known as tread milling or actin recycling (Figure 10), actin monomers severed by cofilin travel with the help of ATP binding, from the actin filament minus-end to the plus-end where they are reassembled.

1.6.2 LIM-KINASE AND COFILIN

The actin de-polymerization factor cofilin is essential for axon growth in *Drosophila* neurons. Rho GTPases are linked to the actin cytoskeletal machinery during axon growth and cofilin regulates this critical step (Ng and Luo, 2004). Interestingly, Meber and Bamburg (2000) provided direct evidence showing that increased cofilin (actin depolymerizing factor/cofilin (ADF)) activity promotes process extension and neurite outgrowth. They did this by overexpressing ADF/cofilin and then measuring the neurite lengths (Meberg and Bamburg, 2000). Cofilin is deactivated by LIMK (Figure 9). Through phosphorylation of LIMK, cofilin is switched off, and thus, does not promote actin de-polymerization anymore. In humans the LIM-kinase protein family has two highly related members, LIMK1 (LIM kinase 1) and LIMK2 (LIM kinase 2), which are encoded by separate genes (Scott and Olson, 2007).

A disease called Williams syndrome has been mapped to a deletion of chromosome 7q11.2, which contains more than 20 genes including LIMK1. This disease is a rare genetic disorder with symptoms like narrow arteries and mild to moderate mental retardation (Meyer-Lindenberg *et al.*, 2006). Although the missing LIMK1 gene can contribute to the development of Williams syndrome, additional genes in this region may play a role and add to the overall spectrum of this disease.

2 MATERIAL METHODS

2.1 Material

2.1.1 FLY STRAINS

Table 1- List of flies

GENOTYPE	COMMENT	SOURCE	REFERENCE
loe	*loechrig* mutant	D. Kretzschmar	(Deák et al., 1997)
w* sn1; e1	white, ebony mutant	Bloomington SC	1870
WT- Canton S	Wild-type flies n special	stock collection	(Lindsley and Zimm, 1992)
w^{1118}	white mutant	stock collection	(Lindsley and Zimm, 1992)
w/FM7a; D/TM3Sb	White mutant balanced on the 1st and 3rd chromosome	stock collection	
w; CyO/Sco; D/TM3Sb	White mutant balanced on the 2nd and 3rd chromosome	stock collection	
w; (elav-GAL4/TM3Sb)}	neuronal GAL4 expression line	Bloomington SC	
w; (loco-GAL4/TM3Sb)	glial GAL4 expression line	C. Klämbt	
w; (GMR-GAL4/CyO)	retinal GAL4 expression line	M. Freeman	
w; (actin-GAL4/TM3Sb)	ubiquitous GAL4 expression line	Bloomington SC	
elavAppl-Gal4/TM3Sb	Gal4 expression under the control of the elav and Appl promotor	stock collection	
[1] v[1]; P{y[+t7.7] v[+t1.8]=TRiP.JF02060}attP2	Expresses dsRNA for RNAi of SNF4Agamma (FBgn0025803) under UAS control,	Bloomington SC	26291

	TRiP		
y1 w67c23; P{lacW}Fppsk06103/CyO	FPPS- insertion on II-chromosom	Bloomington SC	10604
y1 w*; P{UAS-ssh.N}30	Expresses wild type SSH under UAS control	Bloomington SC	9112
P{UAS-tsr.N}1.7, y1 w*; Pin1/CyO	Expresses wild type tsr (cofilin) under UAS control	Bloomington SC	9234
P{UAS-tsr.S3A}4.1, y1 w*; Pin1/CyO	Expresses a constitutively-active tsr (cofilin) protein under UAS control	Bloomington SC	9236
y1 w*; P{FRT(whs)}G13 P{A92}tsr1/CyO, P{sevRas1.V12}FK1	tsr - insertion on II-chromosom	Bloomington SC	9107
y1 w*; Rho172O/CyO, P{sevRas1.V12}FK1	Rho1 – insertion on II-chromosom	Bloomington SC	7325
y1 w*; Rho172F/CyO	Rho1 – insertion on II-chromosom	Bloomington SC	7326

2.1.2 ANTIBODIES

2.1.2.1 Primary

Table 2- List of primary antibodies

ANTIBODY	ANTIGEN	ORGANISM	SOURCE	Dilution WB
22C10	*Drosophila* futsch	mouse	DSHB	1:200
E7	tubulin	mouse	DSHB	1:200
JLA 20	actin	mouse	DSHB	1:200
α-G-actin	actin	rabbit	Cytoskeleton	1:200
ADL101	*Drosophila* lamin	mouse	DSHB	1:200
farnesyl	farnesyl	rabbit	abcam	1:200
α-p-cofilin	28phosphor-cofilin	rabbit	(Jovceva *et al.*, 2007)	1:300
PD-190	*Drosophila* Rho	mouse	DSHB	1:100
α-TSR	*Drosophila* cofilin	rabbit	(Niwa *et al.*, 2002)	1:500
α-SSH	*Drosophila* slingshot	rabbit	(Niwa *et al.*, 2002)	1:2000
NC 82	bruchpilot	mouse	Buchner	1:10
α-sws	swiss cheese	rabbit		1:500
AB49	cysteine string protein	mouse	DSHB	1:500

2.1.2.2 Secondary

Table 3- List of secondary antibodies

ANTIBODY	ANTIGEN	ORGANISM	SOURCE
α- mouse	mouse IgG	Peroxidase-conjugated sheep	Jackson
α- rabbit	rabbit Ig	Peroxidase-conjugated donkey	Jackson
α- chicken	rabbit IgY (IgG)	Peroxidase-conjugated donkey	Jackson

2.1.3 LADDERS AND ENZYMES

Table 4- List of ladders

LADDER	SOURCE
1 kb Ladder	NEB
2 Log DNA Ladder (0.1-10kb)	NEB
Prestained Protein Molecular Weight Marker	Fermentas

2.1.4 KITS

Table 5- List of kits

NAME	SOURCE
G-actin/Factin Assay Kit	Cytoskeleton
GenJET PCR Purification Kit	Fermentas
GenJET Plasmid Miniprep Kit	Fermentas
QIAquick Gel Extraction Kit	QUIAGEN

2.2 Methods

2.2.1 WORKING WITH FLIES

2.2.1.1 Fly keeping

Stocks were kept at 18°C on standard corn meal fly food. The generation time for flies grown at 18°C is 19 days and at 25°C 10 days. Crosses and aging experiments were performed on standard food (except feeding experiments; see 3.3.2) at 25°C.

2.2.1.2 Paraffin sections

Paraffin sections of fly heads were used to determine and compare the level of vacuolization of various fly strains. Photoshop was used to calculated total pixel number (converted into µm2) of the vacuoles in the lamina, medulla, lobula, and lobula plate of each brain hemisphere.

2.2.1.2.1 *Preparing collars:*
- thread the flies in the collar
- fixation in Carnoy's fix 4h
- dehydration in 99% Ethanol 2 x 30 min following 100% Ethanol 45 min
- in Methylbenzoat p.A room temperature, 4 h or over night.
- 1:1 Methylbenzoat- Paraffin mix at 60°C for1h
- Paraffin washes at 60°C, 6 x 30min

Casting collars:- embedding collars in suitable mold in hot paraffin

Cutting heads: fix the paraffin blocks on warm metal block by melting briefly
- carve into 5µm thick slices
- slices are placed on a poly-L- lysine coated slide

- dry at room temperature over night
- removing the paraffin with safe clear 2x 30min
- embedding in mounting medium

2.2.1.3 Fast phototaxis assay

For the fast phototaxis assay, a countercurrent apparatus (first described by (Benzer, 1967)) is used to rate the speed of flies walking toward the light. Five repeated cycles (6 s each) were used to determine the performance index. The performance index is dependent on the number of flies that stayed in the first tube (performance index- 0) or passed to any of the following five vials (reaching the last vial- performance index- 100). At least eight independent tests with groups of 5-15 flies were performed for each genotype, gender. A detailed description of the conditions can be found in (Strauss and Heisenberg, 1993).

2.2.2 WORKING WITH PROTEINS

2.2.2.1 Western Blot

Prearrangement:
- decapitating flies
- grind heads (10 per tube) in 35µl loading buffer
- heat in temp.block 95°C for 5 min
- loading the SDS-Gel
- the gel runs in LAEMMLI buffer at 150V for 60 min

Blotting:
- soak filter paper (whatman) and nitrocellulose or PDVF membrane in transfer buffer
- prepare the gel sandwich for the wet transfer electrophoretic transfer cell (BIORAD):

  3 x Filter paper- 1 x Nitrocellulose membrane -SDS- PAGE- Gel 3 x Filter paper
  - blotting 1h 100V at RT or over night 30V 4°C

Blocking:
- block membrane in 5% dry milk/TBST(milk buffer) 1h
- incubate in primary antibody in milk buffer at least 1h
- rinse in TBST 3 x 10 min
- incubate in secondary antibody in milk buffer at least 1h
- rinse in TBST 3 x 10 min

Displaying:
- mix Visualizer Reagent 1 and 2 in appropriate ratio
- overlay blot in detection solution for 1-5 min

Detection:
Expose leave the blot in the 10sec- 1h the Biospectrum Imaging system (UVP)

2.2.2.2 Visualizing F-actin and G-actin levels

To visualize the amount of filamentous and globular actin in the fly head, the in vivo assay kit from Cytoskeleton was used. The assay principle is that cells are lysed in a detergent-based lysis buffer that stabilizes and maintains the G– and F– forms of cellular actin. The buffer solubilizes G-actin but will not solubilize F-actin. A centrifugation step pellets the F-actin and leaves the G-actin in the supernatant. Samples of supernatant and pellet are run in an SDS-PAGE system and actin is quantitated by western blot analysis.

For each sample 10 fly heads were used and prepared as described in the Cytoskeleton protocol.

For more detailed information look at the Cytoskeleton Cat. # BK037 user book.

2.2.3 IMMUNOPRECIPITATION AND SUBCELLULAR FRACTIONATION

For immunoprecipitations approximately 500 heads were homogenized and Rho1 precipitated following the protocol in (Swanson et al., 2005, #2096), using the anti-Rho antibody and Gammabind bead TM plus Sepharose (Amersham Biosciences) beads (Vector laboratories).

Membrane and cytosolic fractions were prepared from the different genotypes following the protocol of Orgad et al. (Orgad et al., 1987, #2100). Approximately 500 heads were used for each preparation, protein amounts determined by Bradford assays (Bradford, 1976, #2143) and 10μg total protein loaded per lane.

2.2.4 WORKING WITH PRIMARY CELLS

For the cell culture experiments 3rd instar larvae brain cells were used.
- Prepare enzyme solution: collagenase 1mg in 4ml Rinaldini solution.
- Washing: put larvae in ethanol bath and afterwards two times in PBS
- Dissect brains out of larvae in the PBS and transfer brains into enzyme solution
- After all brains are in enzyme use forces to rip the brain into small pieces
- Transfer all pieces into eppendorf tube with 0.5-1ml of enzyme
- Let brains at room temperature for 60 to 90 min depending on the freshness of the enzyme
- Spin down the tubes for 6 min max speed, you should see a small pellet
- Remove enzyme solution and re-suspend in media or Schneider's 1ml. Using a fire polished glass pipette triturate the pellet
- Spin down the tubes for 6 min max speed. Remove media and resuspend in fresh full media 200μl.

- Pipette cells and media to distribute cells evenly and then plate 125µl of mixture into each culture dish. Make sure small dishes are in larger dish to keep them from drying out in the incubator.
- Incubate 2h to overnight so cells can adhere to dish.
- The next day flood dish with dish media 2ml. Be gentle when adding media so you do not wash the cells of the dish
  - Feed cells with new media every two to three days.

2.2.4.1 Measuring neurite length and movement of mitochondria

Neuronal cell cultures were prepared from 3rd instar larvae as described by (Kraft *et al.*, 1998). To determine neurite number and length photographs were taken after 24h and 48h in culture without knowing the genotype, using a Leica inverted microscope. The number of neurites each cell had extended was counted and the length of the longest neurite measured in pixels using ImageJ and converted into µm before the genotype was determined. Measurements of mitochondrial movements were performed on 24h old primary cultures, using green Mitotracker CM-H2XRos (Molecular Probes, Eugene, USA). Cells were stained for 10 min and then observed with an inverted microscope. Images were taken every 2 seconds for 4.5 minutes. To perform an analysis of mitochondrial movement, we used the tracking function in Metamorph Universal Imaging and created tables with the amount of pixels each mitochondrion moved after 2 seconds. The average distance traveled in the 2 second intervals was determined and converted into velocity

3 RESULTS

The experiments described in this thesis where performed using the very helpful and straightforward model, *Drosophila melanogaster*, to determine the functions of AMPK (AMP- activated protein kinase) and its downstream targets.

As shown by Tschäpe *et al.*, 2002, flies homozygous for *loe* and heterozygous for lethal alleles of HMGR, which in *Drosophila* is encoded by the *columbus* gene, show a suppression of the degenerative phenotype compared to *loe* alone. These experiments confirmed that the inhibitory function of AMPK on HMGR is conserved in flies and that changes in the activity of HMGR play a role in the observed degenerative phenotype. HMGR is a key factor in cholesterol synthesis but also in isoprenoid synthesis, a pathway conserved in *Drosophila* (Figure 8).

AMPK negatively regulates HMGR and the isoprenoid pathway, therefore one would expect more farnesylation, when AMPK is impaired by the *loe* mutation. It was previously shown that FPPS (*Farnesyl- pyrophosphate-synthase*), which is essential for farnesylation, influences the *loe* phenotype and a mutation in this gene has an suppressing effect on the vacuolization in *loe* (Kadshojan, 2006). This suggested that targets of the isoprenoid pathway, like GTPases, may interfere with the *loe* phenotype. In order to reveal a possible correlation between GTPases and LOE the following experiments where performed.

3.1 Interactions with RHO and ROK

To determine whether the Rho1 pathway is involved in neuronal survival in *loe*, double mutants that had mutations in genes that are involved in the Rho1 pathway and *loe* were generated, aged and tested for the amount of neurodegeneration within the adult brain. Furthermore, immunoprecipitation and Western Blot analysis was used to quantify the amount of isoprenylated Rho1.

3.1.1 LOE SHOWS MORE ISOPRENYLATION OF RHO1

To confirm the hypothesis that *loe* affects the levels of prenylated Rho1, Rho1 was immunoprecipitated from head lysates using a α-Rho1 antibody followed by Western Blots with a α-farnesyl antibody.

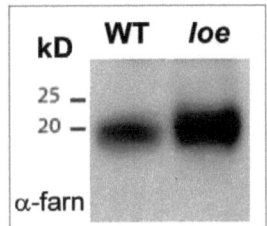 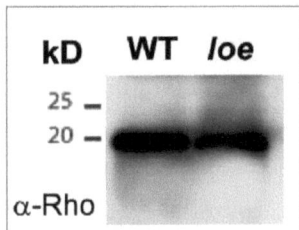

Figure 11 – *loe* affects the levels of farnesylation of Rho1
Immunoprecipitation from head lysates using anti-Rho1 probed with an α-farnesyl antibody (left panel) shows an increase in farnesylated Rho1 in *loe*. The identity of this band with Rho1 was confirmed by reprobing with α-Rho1 (right panel).

As shown in Figure 11, *loe* flies showed increased levels of prenylated Rho1 (left panel) when compared to wild type. Surprisingly, the western blot also showed that the total levels of Rho1 are reduced (right panel). This confirms the hypothesis that the isoprenoid pathway is upregulated in *loe* and leads to hyperfarnesylation of Rho1 in *loe*. This suggests that interfering with the prenylation of Rho1 might activate a regulatory process that reduces Rho1 transcription or protein stability.

3.1.2 MORE RHO IS LOCATED AT THE MEMBRANE IN *LOE*

The previous experiment demonstrated that *loe* has more farnesylated Rho1, which suggests that because the attachment of isoprenyl moieties facilitates membrane association of small GTPases, the majority of Rho1 in *loe* should be located at the membrane. To address this, membrane and cytosolic fraction were prepared from *loe* and control flies (as a control for the separation of membrane and cytosol fractions two antibodies (CSP and tubulin) were used as markers).

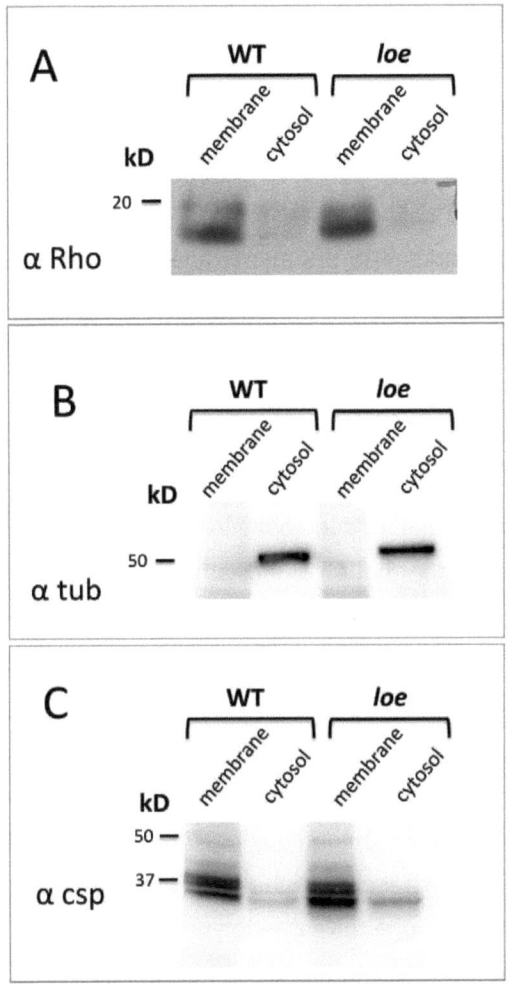

Figure 12 – More Rho is found at the membrane in *loe*
Western Blot using membrane and cytosolic fractions shows an increase of membrane-associated Rho1 and a decrease of cytosolic Rho1 in *loe*

(A). α-tubulin antibody (B) serves as cytosolic and α-CSP antibody (C) as membrane marker. 10μg of total protein were loaded in each lane.

Figure 12 C and B shows that CSP, a protein associated with synaptic vesicles, is mostly restricted to the membrane fractions. In contrast, tubulin, a cytosolic protein is almost exclusively found in the cytosol fraction, confirming a separation of membranes from the cytosol fractions. Comparing the levels of Rho1 in Western Blots, *loe* flies show a clear reduction of cytosolic Rho1 compared to control whereas they show a small increase in membrane associated Rho1 (Figure 12A). Comparing the weak increase in membrane associated Rho1 with the strong increase in farnesylated Rho1 in *loe* (as shown in Figure 11) suggests that not all of the farnesylated Rho1 is actually attached to membranes in *loe*. Together these experiments confirm the hypothesis that *loe* mutant flies have an increase in farnesylated, and although weak, membrane- associated Rho1.

At this point it is worth mentioning an experiment, that was performed by P. Mani (Cook et al. 2012) to investigate whether *loe* genetically interacts with mutations in Rho1. Combining one copy of Rho^{72F}, an allele that deletes part of the coding region of Rho1, including the translation start site, with homozygous *loe* reduced the neurodegenerative phenotype almost by half. This result shows that the protein Rho1 is affecting the vacuolization and that down regulating it, is beneficial for the *loe* phenotype. Therefore the question arose whether increased Rho1 levels or activity can aggravate the degeneration in *loe*. For this purpose, *loe* flies with an UAS-RHO construct ($Rho1^{V14}$) driven by a pan-neuronal *Appl*-GAL4 promoter where examined. Rho1^{V14} has a p-element insertion resulting in constitutively active Rho1. With a 70% increase in vacuolization, the expected enhancement of the neurodegenerative phenotype in a *loe* fly that expresses constitutively active Rho was verified.

In addition western blots with an anti-Rho1 antibody where used to confirm that the genetic manipulations did affect the levels of Rho1. At this point it has not been addressed whether an enhancement can also be achieved by increasing the levels of wild type Rho1.

3.1.3 THE INFLUENCE OF ROK1 ON THE NEURODEGENERATIVE PHENOTYPE IN *LOE*

After finding out that the amount of isoprenylated Rho was changed in *loe* and Rho1 genetically interacted with *loe*, looking at the Rho Kinase, which is a downstream target of Rho1 (Figure 9), was the next step.

For this experiment, and for all the following genetic interactions, *loe* flies carrying the additional mutation were compared to *loe* flies from the same cross without the additional mutation, to minimize genetic background effects. Analyzing head sections from five and ten day old female flies homozygous for *loe* with age-matched flies that, in addition, carry one mutant copy of the *rok^1* allele, revealed a significant suppression (from 376 to 226µm^2 holes in the brain) of the degenerative phenotype in 10 day old flies (Figure 13). A small reduction of the area of vacuoles was also found in 5 day old flies, although it was not significant at this age. Nevertheless this shows that reducing the levels of *rok^1* can partially counteract the increased activity of the Rho pathway. The heterozygous *rok^1* mutation alone did not exhibit holes.

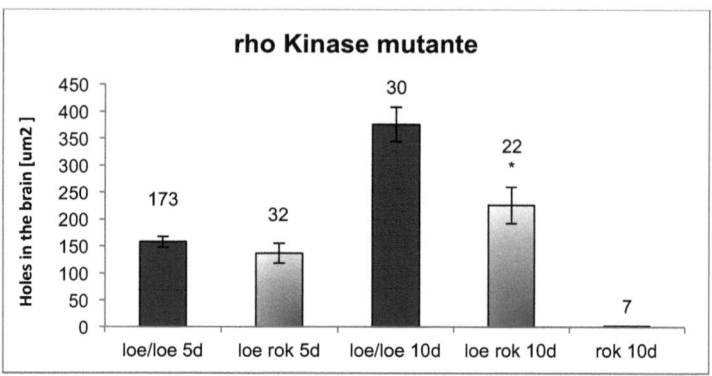

Figure 13 – A Rho- Kinase mutant decreases neurodegeneration
10 day old *loe rok^1* mutants show a significant improvement (about 30%) in neurodegeneration. The number of measured optical lobes is given above each bar. SEMs are indicated

3.2 Effect of the Proteins LIMK and Slingshot on *loe*

To investigate the influence of further downstream proteins of the Rho pathway, like LIM-Kinase (cofilin-inactivator) and Slingshot (SSH; cofilin-activator), flies where generated that overexpress LIMK or SSH in in the *loe* background. The Gal4-UAS-System, described in Figure 14, was used for these overexpression experiments.

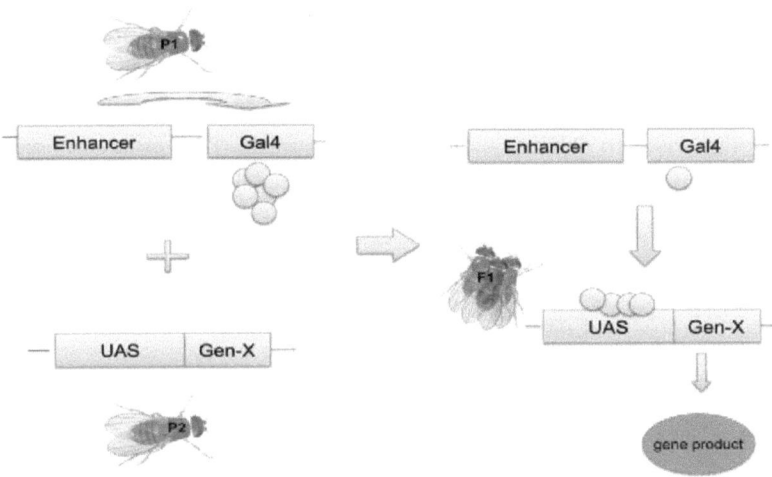

Figure 14 – Schematic of the GAL4- UAS system in *Drosophila melanogaster*

The Gal4-UAS-System is a unique tool that can be used in *Drosophila melanogaster* to express a gene of interested in a desired location in the fly body and was developed by {Brand and Perrimon, 1993, #2493}. This system allows analysis of function of certain gene products. The UAS (Upstream Activating Sequence) - and the GAL4 line must be crossed to obtain the fly with both elements. The heterozygote F1 flies contain the Gal4 driver and responder gene. The driver does not activate native *Drosophila* genes. In the F1 generation the Gal4 protein product binds directly to the UAS element. Subsequently the Gal4- UAS complex drives the expression of the gene located downstream of the UAS sequence. The responder gene only gets expressed in presence of Gal4. This method allows expression of mutated genes in a tissue specific manner.

The UAS-LIMK or UAS-SSH constructs where driven with Appl-Gal4, a pan neuronal driver to achieve expression in neurons. To quantify the amount of neurodegeneration in LIMK or SSH overexpression *loe* flies, the same method as mentioned in the previous experiment 3.1.3 was used.

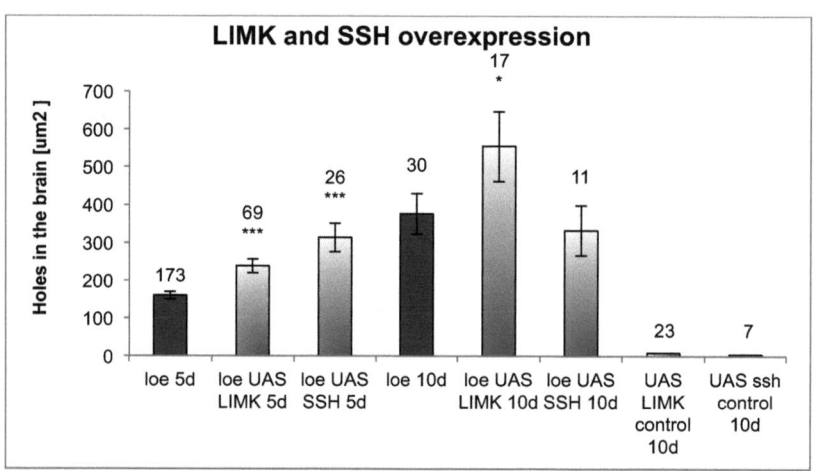

Figure 15 – Increased activation of LIMK or SSH results in enhanced neurodegeneration in *loe*

5 and 10 day old LIMK overexpression flies show an enhanced neurodegenerative phenotype with an increase of about 50% in both ages. The Slingshot (SSH) over-expression fly only enhances the 5 day old *loe* phenotype. The number of measured optical lobes is given above each bar. SEMs are indicated

As shown in Figure 15, the neurodegeneration is significantly increased in *loe* flies that overexpress LIMK. After 5 days the area of holes increased significantly from 159 to 238μm^2 and after 10 days from 376 to 554μm^2.

Because after expressing more LIMK, which should lead to even more active LIMK, in the *loe* background, the vacuolization is significantly increased. This further confirms the hypothesis that the up-regulated Rho1 pathway is connected with the neurodegenerative phenotype in *loe*. More LIMK means more phosphorylated cofilin, which would result in more inactive cofilin. Therefore the neurodegeneration in *loe* could be caused by higher levels of inactive cofilin or reduced levels of active cofilin because it is inactivated by phosphorylation.

However, looking at the SSH results shows that this is not the case. Slingshot dephosphorylates cofilin, which is the opposing action of what LIMK does, nevertheless overexpressing SSH did not improve the *loe* phenotype - it made it worse. Currently it is unknown why both have the same effect. However the slingshot overexpression in the *loe* background only affected the neurodegenerative phenotype after 5 days, where holes increase from 159 to 313μm^2. After 10 days of aging the SSH overexpression shows neither enhancement nor suppression. This suggests that at this age, a compensatory mechanism might start taking place that prevents the effects of SSH overexpression in 10 day old flies.

Overexpression of LIMK or SSH by itself show no phenotype, which demonstrate that LIMK and SSH enhance neuronal death only in connection with the *loe* mutation.

3.3 The Protein Cofilin

Since the previous results demonstrate the involvement of LIMK and SSH in the degenerative phenotype of *loe*, the next experiment is focused on the protein cofilin. Cofilin is a key protein in the RHO pathway, which is phosphorylated by LIMK and dephosphorylated by SSH. As an actin depolimerization factor, cofilin plays an important role in actin dynamics.

3.3.1 THE INFLUENCE OF COFILIN ON THE NEURODEGENERATIVE PHENOTYPE IN *LOE*

To prove that cofilin, which in *Drosophila* is encoded by the *twinstar* gene, is involved with the *loe* phenotype, neurodegenerative analysis like in 3.1.3 and 3.2, and western blotting were used.

After seeing the enhancement of degeneration in the UAS-LIMK *loe* and UAS-SSH *loe* flies the question was, whether the phenotype could be due to increased levels of cofilin, specifically an accumulation of phospho-cofilin. It was therefore investigated whether *loe* flies with a mutation in twinstar show changes in the degenerative phenotype.

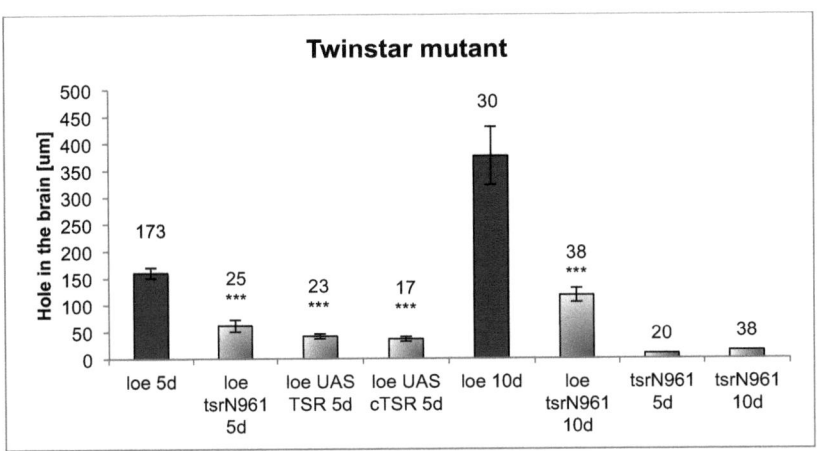

Figure 16 – Effects of TSR on *loe* neurodegeneration
Interestingly both, a mutant in *tsr* and TSR overexpression suppress the area of holes. The number of measured optical lobes is given above each bar. SEMs are indicated.

Loe flies with a heterozygous twinstar mutation should have a lower level of total cofilin, which should then also decrease phospho-cofilin and reduce the vacuolization. And indeed, a significant suppression was found in $tsr^{N961}loe$. After 5 days the area of holes was reduced from 159 to 60μm² in $tsr^{N961}loe$. Also the phenotype significantly improved in the $tsr^{N961}loe$ mutant fly after 10 days. By itself, tsr^{N961} showed no neurodegeneration after 5 or 10 days (Figure 16).

In contrast to the *tsr* mutation an overexpression of twinstar should show an enhanced phenotype because of the higher level of cofilin and inactive cofilin mimicking the overexpression LIMK phenotype. Surprisingly, the area of holes in TSR *loe* was reduced from 159 to 41μm^2. These results show that a change of cofilin levels in either direction improves the neurodegenerative phenotype in *loe*. To address whether changing the levels specifically of active, dephosphorylated cofilin, a constitutively active form of TSR, was expressed. Interestingly, this also reduced the degeneration (cTSR *loe* flies show reduction from 159 to 35μm^2) although overexpression of SSH, which should have the same effect, enhanced the phenotype.

3.3.2 UP-REGULATION OF INACTIVE COFILIN IN *LOE*

Due to the up-regulation of LIM-Kinase in *loe* flies, the amount of phosphorylated cofilin should be increased as well. In order to test that, Western blotting was used to compare the level of phospho-cofilin in *loe* and wild-type flies (Figure 17).

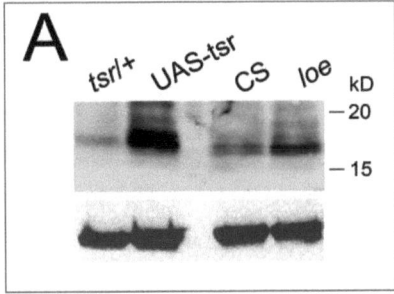 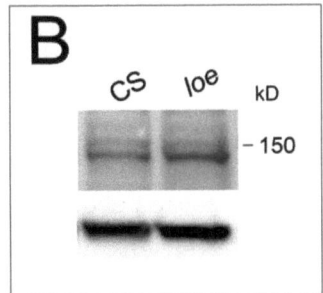

Figure 17 – Increased level of P-Cofilin and SSH in *loe*
Western Blot probed with a α- p-cofilin antibody reveals an increase in p-cofilin in *loe* (A-last lane). *Twinstar* mutants (first lane) and twinstar overexpression (second lane) show the expected decrease and increase of p-cofilin. A western blot using α- ssh antibody, shows that the *loe* fly has an increase of about 20 % in slingshot expression (B-upper panel). As loading control α-tubulin was used in A and B (lower panels).

Figure 17 A showed that, as expected, the level of p-cofilin (inactive cofilin) is increased in the *loe* fly compared to controls. This confirmed that the hyperactivity of the Rho1 pathway indeed results in increased levels of the downstream target p-cofilin. *Twinstar* mutant and overexpression confirm that the α-p-cofilin antibody is specific and that the genetic manipulations of *tsr* in 3.3.1 had the expected effects on p-cofilin levels. Unfortunately, the overall levels of cofilin could not be determined, because the obtained antibody did not work in western blotting. After demonstrating that the p-cofilin level is increased in *loe*, it was interesting to find out if this increase affects the slingshot expression.

The increased phospho-cofilin level found in *loe*, could result in an up regulation of SSH to remove the p-cofilin. And in fact, as shown in Figure 17B, the SSH expression is increased about 20%, which could be a reaction to the increased inactive cofilin level. Also the unexpected results of the genetic interactions (Figure 15) could be explained after finding out, if the level of SSH is already increased in *loe*. The already elevated level of SSH in *loe* might explain why additional expression of SSH is not beneficial but even enhances the neurodegenerative phenotype.

3.4 The Role of Actin

Because of the clear association of cofilin with the *loe* mutation and because cofilin promotes actin disassembly, experiments to detect actin dynamics were performed. A higher amount of inactive cofilin should lead to less 50epolimerization of actin in *loe* (look at Rho-pathway Figure 9). Which means an increased presence of filamentous actin and therefore decreased levels of free actin monomers (G-actin).

To find out if actin turnover and polymerization is actually affected, a G-actin/F-actin assay was performed that indicates the ratio of monomeric depolymerized globular actin, which is found in the supernatant versus polymerized filamentous actin, which is found in the pellet.

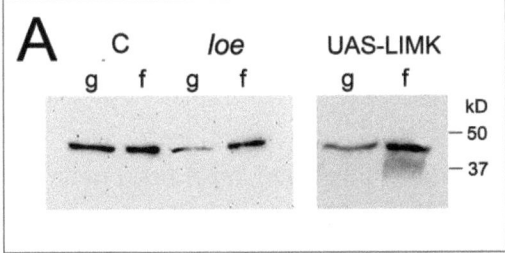 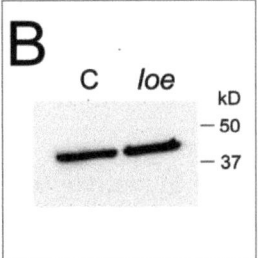

Figure 18 – Loe shows increased level of fibrilar actin
Western blot with α-actin antibody. Left side- the first two lanes show that in wild type the amount of both forms of actin are found at equal levels. *Loe* shows a higher amount of insoluble F-actin Right side- LIMK overexpression, as expected, also shows more insoluble F-actin compared to G-actin. (B) Blot with α-actin antibody shows no difference of the total actin concentration in wild type and *loe* g: Supernatant fraction represents globular actin. f: Pellet fraction represents filamentous actin

As predicted, the balance between free actin and filamentous actin is disrupted in *loe* (Figure 18A), with more filamentous actin than globular actin in the mutant. The same phenotype can be observed in a fly that expresses more LIMK and is serving as a positive control and confirms that a hyperactivation of the Rho pathway does indeed lead to more filamentous actin. The total actin concentration in *loe* compared to wild type is unchanged supporting the hypothesis that *loe* leads to a change in actin dynamics and does not have effects on the levels of actin per se.

3.5 *loe* neurons show a change in neurite outgrowth and organelle transport

Due to the role of actin in neuronal outgrowth, larval brain neurons were dissected and cultivated for 24 hours or 48 hours and analyzed for outgrowth phenotypes. Although *loe* mutant flies do not exhibit detectable defects in the development of the brain (Tschape *et al.*, 2002), these primary neuronal cell cultures could reveal more subtle effects that might be visible in these isolated cells. To determine if the outgrowth of neurites is disturbed, measurements of the longest process where taken and compared to wild type neurites. In addition the number of neurites formed by each cell were determined (see 3.5.1). To investigate whether the transport of proteins and organelles through the axons and dendrites might be affected, the speed of the mitochondria was recorded. (see 3.5.2)

3.5.1 OUTGROWTH OF NEURITES IS AFFECTED IN *LOE*

The genetic interaction studies shown in Figure 13, Figure 15 and Figure 16, strongly suggest that LOE interferes with the dynamics of the actin network. Due to the well-known role of actin dynamics during neuronal outgrowth, the following experiments were performed to elucidate whether the *loe* mutation affects neural outgrowth and/or morphology. As mentioned before *loe* does not visibly affect brain development. To investigate if larval neurons show a phenotype, processes of primary neurons from 3instar larva were measured after being cultured for one or two days, (Figure 19). Testing the cells after 24 hours, helped to determine the initial outgrowth of neurites, whereas looking after 48 hours should reveal if the neurites can maintain the outgrowth.

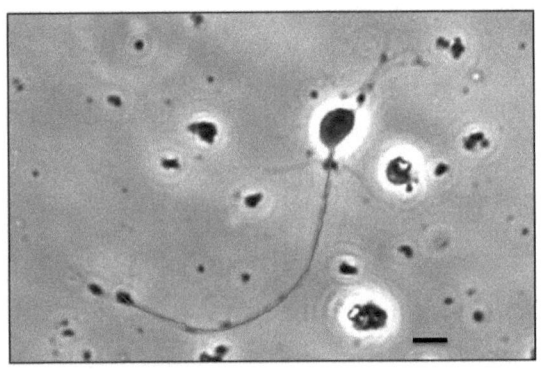

Figure 19 – *loe* neuron cultured for twenty-four hours
The longest neurite (highlighted in red) was measured and compared to control neurons. Scale bar= 2μm.

Whereas no changes in neurite numbers or branching patterns were found between wild type and *loe* neurons grown for 24h (Figure 20, lower graph), measurements of the longest neurite of each cell revealed a significant difference in length. Surprisingly the neurites were significantly longer in *loe* with 22.4 μm in wild type and 34.4 μm in *loe* ($p<0.001$; Figure 20, upper graphs).

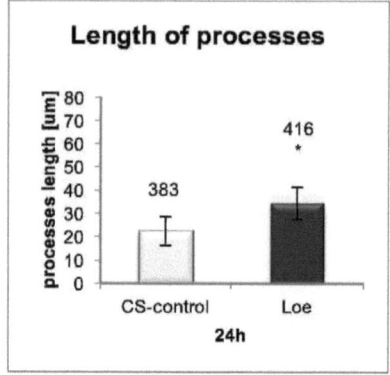
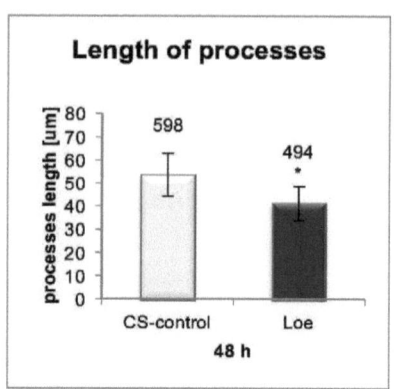
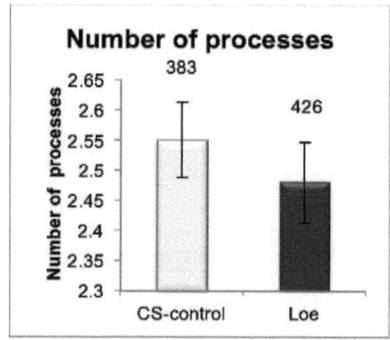

Figure 20 – Neurite length of larval brain neurons

Length of the longest processes of each cell in *loe* and wild type neurons. *Loe* processes seem to grow faster in the first 24h. After 48 h in culture the processes of loe neurons are shorter than wild type.. Lower graph- The number of processes is unchanged. The number of measured cells is given above each bar SEMs are indicated.

These results show that the initial outgrowth is occurring even faster in *loe*. To determine whether *loe* neurons can maintain neurite growth, 48h old cultures were measured, which revealed the opposite result. *Loe* processes are significantly shorter, with a length of 41.2μm compared to wild type with 53.4 μm. This result suggests that the increased activation of Rho1 in *loe* and the resulting LIMK activity and inhibition of tsr/cofilin initially promotes outgrowth, possibly by a stabilization of actin filaments. However *loe* cells appear not to be able to sustain this increased outgrowth and are even not able to maintain normal length over time.

The loe mutation does not affect the amount of processes of each cell. The average number is a little decrease however not significantly changed.

3.5.2 SLOWER TRANSPORT THROUGH NEURITES IN LOE

Mitochondrial defects are key features of chronic neurodegenerative diseases. To determine, whether change in the actin cytoskeleton might have effects on organelle movement through the processes, mitochondria were visualized and tracked as described in 2.2.3.1.

Figure 21 shows a neuron that was treated with green Mitotracker CM-H2Xros to visualize mitochondria. Comparing mitochondria motions in *loe* and wild type neurons after being cultured for 24 hours showed that the speed of mitochondria in *loe* cells is about 30nm/sec slower than in wild type ($p<0.01$).

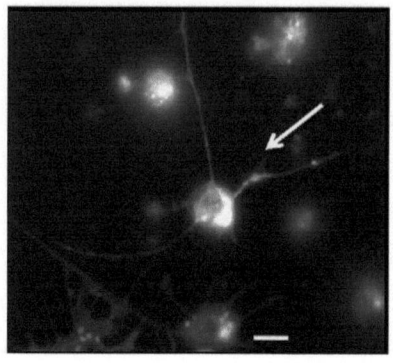 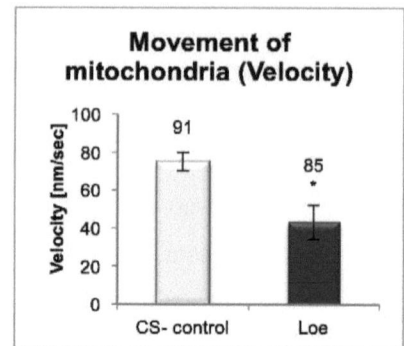

Figure 21 – Velocity of mitochondria and number of processes in neurons
Left side: Neuron from a *loe* larva that was cultured for 24 hours. White arrow points to mitochondrion that moves through a process towards the cell body. Scale bar= 2μm. Right graph- Mitochondria move significantly slower from and towards the cell body in *loe* neurons than in wild type. SEMs are indicated

These results support the hypothesis that interference with the actin dynamics in the *loe* mutant, also leads to changes in the tubulin network and, therefore, a dysfunction in axonal and dendrite transport.

3.6 Behavioral phenotype in *loe*

Many degenerative diseases exhibit behavioral phenotypes before the degenerative phenotype is detectable. Therefore behavior assays have been done with *loe* flies. The neurodegenerative phenotype in *loe* appears a few days after of eclosion. To find out if there is a detectable behavioral phenotype before the neurons undergo necrotic cell death, behavior assays where done in 1-day-old flies.

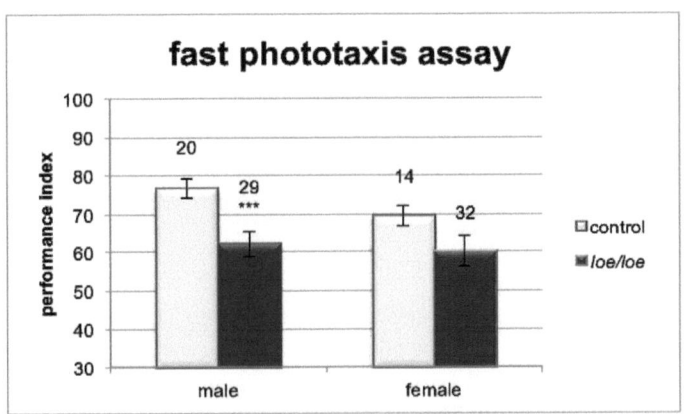

Figure 22 – Phototaxis assay with *loe* flies

Loe males show a highly significant decrease in the performance index. Females also preform worse than the control flies but show no significance. The number of groups tested is given above each bar. SEMs are indicated

To detect behavioral defects a phototaxis assay was used in which males and females were tested separately in groups of 15-20 individuals. Whereas *loe* males have a significant behavioral deficit when compared to control ($p<0.0006$, Figure 22), females showed no significant difference. The neurodegenerative phenotype develops slower in females than males. Accordingly the same could be the case for the behavioral phenotype and therefore the difference is not visible at 1 day.

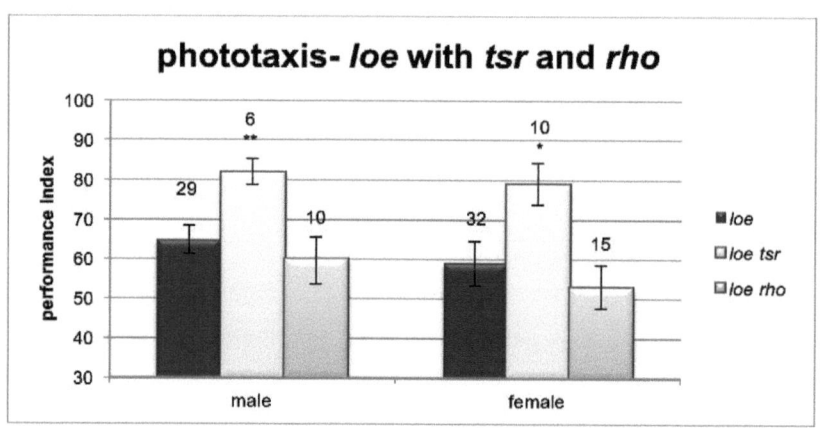

Figure 23 – Phototaxis assay with *loe tsr* and *loe rho* flies
Loe males and females that have a mutation in tsr perform significantly better in the Phototaxis assay than *loe* flies alone. *Loe rho* flies show now difference to *loe*. The number of groups tested is given above each bar. SEMs are indicated

Moreover, *loe* flies were tested that express less cofilin. Both male and female flies performed significantly better ($p<0.009$ male, $p<0.02$ female) in the fast phototaxis assay (Figure 23). This indicates that cofilin plays a relevant role in the *loe* behavior phenotype and by knocking down the cofilin levels in the *loe* fly, the behavioral phenotype improves as does the neurodegenerative phenotype (Figure 16). Unexpectedly the *loe rho* mutants show no significant improvement also *rho* suppressed the neurodegenerative phenotype of *loe*.

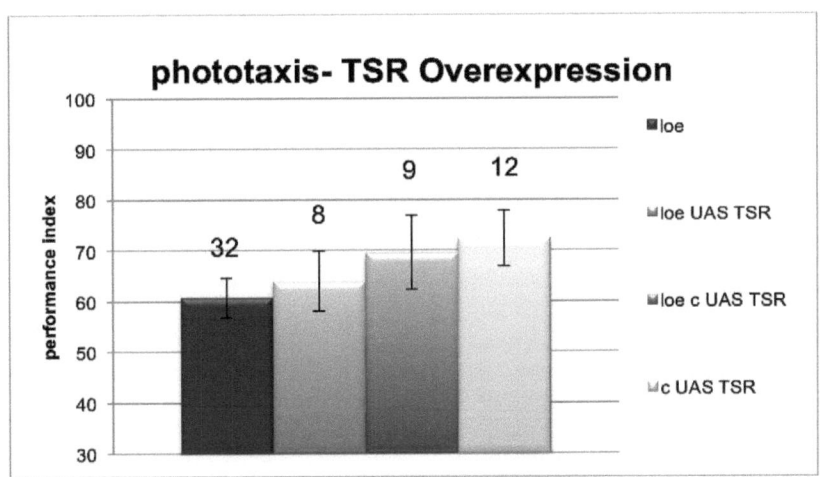

Figure 24 – Phototaxis assay with TSR overexpression
Female *loe* UAS TSR flies do not differ in the behavioral phenotype. The number of groups tested is given above each bar. SEMs are indicated

Given that *loe* flies with a decreased cofilin expression showed a significant improvement in the phototaxis assay, the question was if a overexpression of cofilin would worsen the phenotype. However, the results demonstrate that flies overexpressing (UAS driven with a pan neuronal driver) cofilin in the *loe* background showed no change in the performance index (Figure 24). Overexpression of constitutively active cofilin slightly increased the performance also not statistically significant. Interestingly, overexpression of constitutively active Rho alone resulted in a performance index similar to *loe* suggesting the expression of constitutively active Rho also leads to a behavioral phenotype.

The last and following phototaxis experiment demonstrated that overexpressing slingshot in the *loe* background does not affect the performance index.

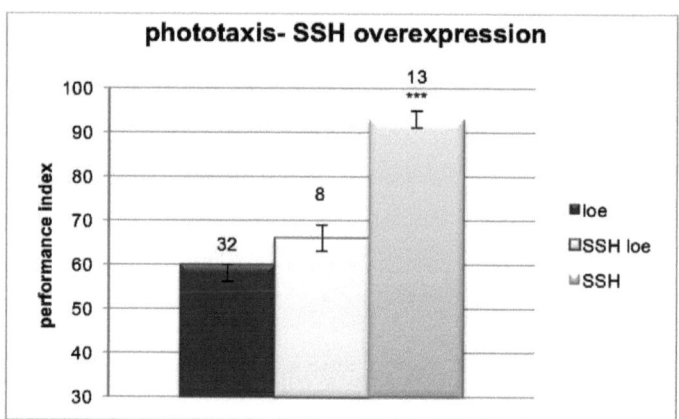

Figure 25 – Phototaxis assay with SSH overexpression
Female UAS SSH *loe* flies do not perform different than *loe* flies. The number of groups tested is given above each bar. SEMs are indicated

Since slingshot overexpression surprisingly enhanced the neurodegenerative phenotype in loe it was interesting to find out if the overexpression affects the behavioral phenotype in the same way. Flies that overexpress (pan neuronal driver) slingshot in the *loe* background showed a slight improvement but no significant change in the performance index. Slingshot overexpression alone did not differ from the wild type performance (Figure 25).

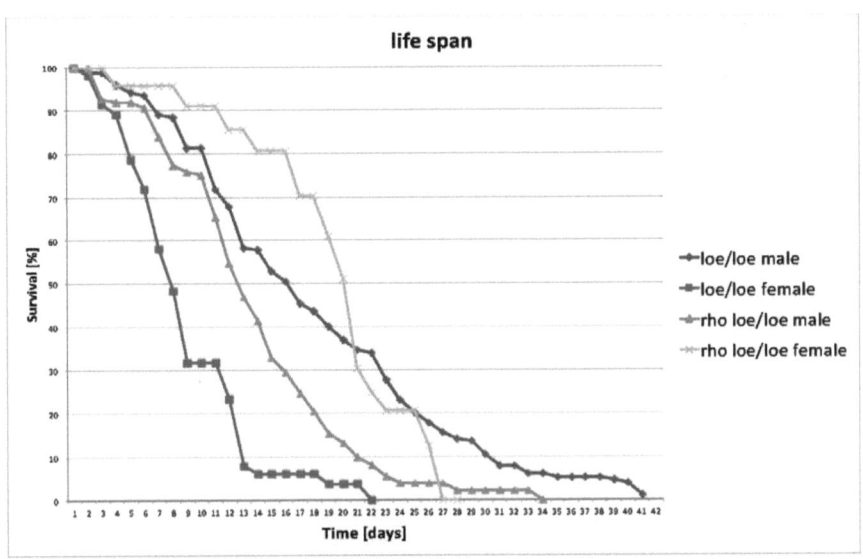

Figure 26 – Life span in *loe* and *rho loe* flies
Loe **females carrying *rho1* mutation live significantly longer than *loe* females alone. Males do not show a significant difference.**

It has been previously shown that the *loe* fly has a significantly shorter life span, than wild type flies. After demonstrating that *rho* influences several *loe* phenotypes, the question is if it also has an affect on the life span. Therefore flies were aged and the time of death was recorded.

The death curve shows that the heterozygote *rho* mutation affected the life span of *loe* flies. The life span of *rho loe* female flies is significantly longer ($p<0.001$ female) than *loe* females. Thus, their life span improved seemingly due to a decreased expression of Rho1 caused by taking out one copy of *rho*. However, an examination of the males showed that the *rho* mutation did not have the same effect on the life span of male *loe* flies. This demonstrates that the *loe* mutation and *rho* mutation disturbs males and females to a different degree. Although beyond the scope of this thesis, this deserves further investigation because it suggests that there are different roles for AMPK in male and female flies.

3.7 AMPK γRNAi shows a degenerative phenotype with various drivers

To find out if a knock down of the AMPK γ subunit transcript has a similar effect as the p-element insertion in AMPK γ subunit (*loe-* mutant), an RNAi line of SNF4A γ was investigated. This RNAi construct specifically affects the loeI transcript. With the UAS-SNF4A γ RNAi (AMPK RNAi) construct, a silencing of the γ subunit in the tissue of interest was possible. The UAS-RNAi line was crossed with several Gal4-drivers and observed for phenotypes in the brain and eye. The most dramatic phenotypes are displayed below.

3.7.1 PHENOTYPE IN UAS-AMPK γ RNAI DRIVEN WITH GMR II – GAL4 AND GMR III – GAL4 AND EY-GAL4

As shown in the figure below, UAS-AMPK RNAi driven by GMR-II and GMR-III (stronger driver) lead to severe neurodegeneration. The GMR-II diver causes vacuoles in the medulla and medulla cortex (A). In contrast, UAS-AMPK RNAi driven with GMR-III leads to complete destruction of the photoreceptor cells (B). Figure C shows that the eye less- GAL4 driver causes, in males, the same phenotype as GMR-II-GAL4 after only one week of aging, compared to 21days with GMR-II. Although the experiments with GMR-III show an important function of AMPK in the eye, it is unknown why its knock down leads to a degeneration in the medulla with GMR-II- and ey- GAL4.

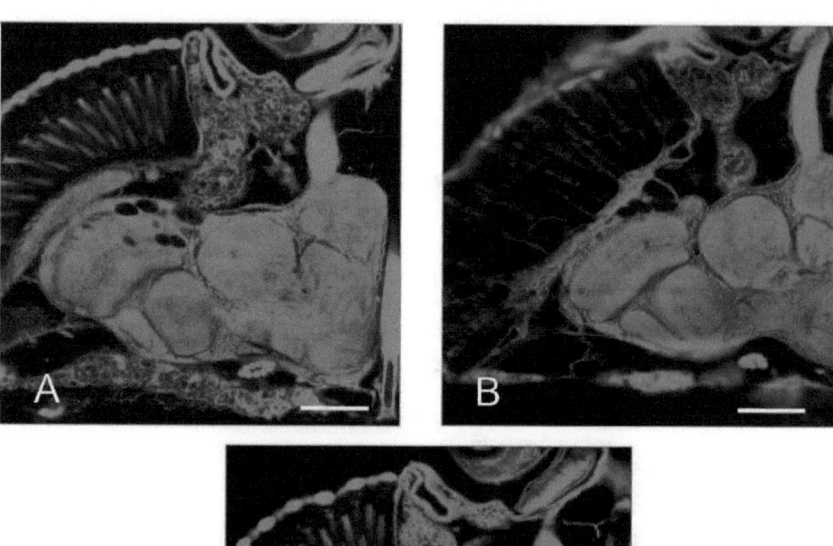
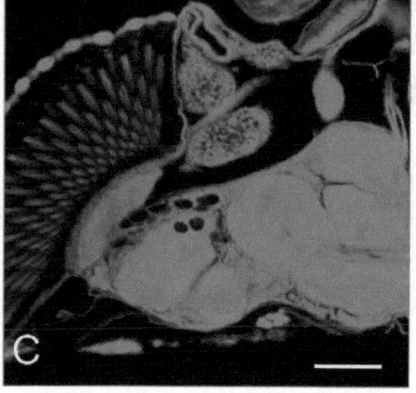

Figure 27 – Cell death in the eye causes by AMPK RNAi
UAS-AMPK RNAi driven with A: GMR-II-GAL4 in 21 day old females and with B: GMR-III-GAL4 21 day old females and with C: Ey-GAL4 in 7day old males causes degeneration. Scale bar= 50µm

3.7.2 PHENOTYPE IN UAS AMPK γ RNAi DRIVEN NEURONAL DRIVERS

Phenotype of AMPK-RNAi with neuronal drivers are shown in the figure below.

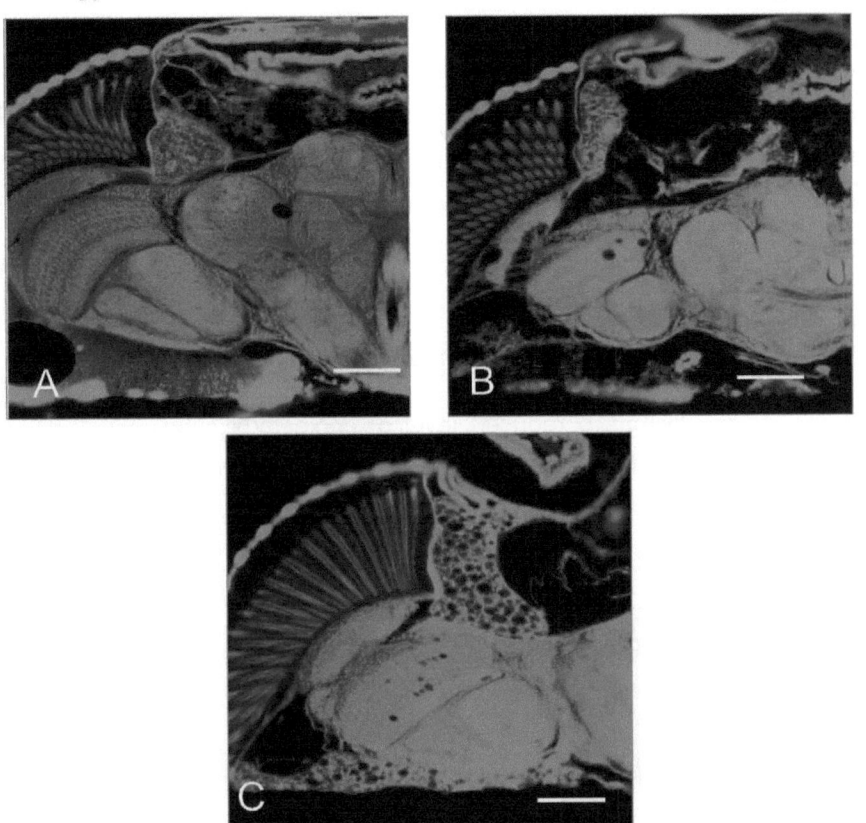

Figure 28 – Degeneration in neurons caused by the AMPK RNAi line

AMPK RNAi driven by A: elav-GAL4 in 42 day old males and B: UAS- dicer Appl- GAL4 in 14 day old females and C: Applelav-GAL4 in 7 day old males Scale bar= 50μm

Figure 28 shows the degeneration that is caused by expression of the AMPK-γ- RNAi line in neurons. After 42 days males that express elav driven AMPK RNAi develop holes in the lateral protocerebrum. Two weeks old AMPK RNAi females that also express dicer via Appl GAL4 show big lesions in the optical lobes. And AMPK RNAi driven with Applelav- GAL4 show cell death in medulla and lobula after only 7 days.

3.8 Cholesterol Influences neurodegeneration

The *loe* fly has a reduced level of cholesterol ester. Nevertheless, the level of its triglycerides, phospholipids and free cholesterol remain steady(Tschäpe *et al.*, 2002). As mentioned before, *Drosophila melanogaster* cannot synthesis cholesterol, yet the cholesterol ester concentration is affected by the mutation in AMPK (Figure 29). Because cholesterol ester is the storage form of cholesterol this could indicate that loe flies do not take up sufficient levels of cholesterol from the standard food.

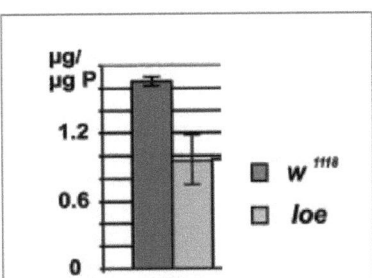

Figure 29 – Cholesterol ester concentration in *loe*
Tschaepe *et al* showed that *loe* has about 60% reduction of cholesterol ester levels. SEMs are indicated

It was therefore tested, whether or not the amount of cholesterol provided in the food can affect the neurodegeneration. For this experiment *loe* and control larva were raised on standard food until pupariation. Fly pupas where transferred to the feeding assay vials that contain tissues soaked in glucose with and without yeast, which ensured that the flies had the special diet from the very beginning of their adult life. They were aged on the special diet for 7 days and then sectioned.

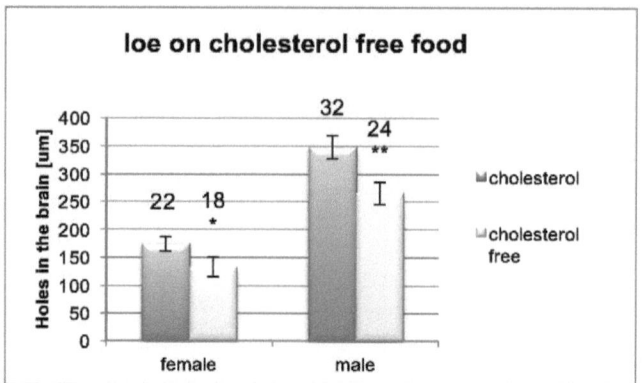

Figure 30 – Feeding assay with cholesterol free food
Male and female flies show significant improvement of the neurodegenerative phenotype on food, that does not contain cholesterol. The number of measured optical lobes is given above each bar. SEMs are indicated

Comparing flies kept on cholesterol containing food with flies that had cholesterol free food showed a decreased area of holes from 163 to 133μm² in females and 348 to 265μm² in males when kept on cholesterol free food (Figure 30).

This surprising result shows that cholesterol enhances the neurodegeneration although the levels of cholesterol are already decreased in *loe* (in the form of cholesterol ester).

4 DISCUSSION

The heterotrimeric protein kinase AMPK, which was first mentioned in 1973 by Carlson and Kim, controls the energy levels of the cell, which requires constant monitoring of the ATP/AMP level. In cases of ATP depletion, its major function is to activate energy-providing mechanisms while inactivating energy-consuming processes. But besides these more general function in metabolism, AMPK has also been shown to regulate protein synthesis, cell growth and cell polarity (Hardie, 2007); (Steinberg and Kemp, 2009a). Therefore, it is expected that AMPK is an evolutionary conserved protein that is expressed in all tissues including the brain (Culmsee et al., 2001), liver, and muscle.

As discussed in the introduction, a mutation in such an important protein as AMPK can lead to many disorders involved in metabolism and of course the nervous system, and therefore more investigation is needed to fully understand and explain how the processes in the cell are affected when AMPK is affected. The insights that have been gained so far are very interesting, encouraging and in some way surprising regarding the effects from just one mutation, in one subunit (p-element insertion in AMPK-*loe* mutation) and how dramatically cell growth, homeostasis and survival is affected.

AMPK has been connected to a rare hereditary disease, Wolf Parkinson White Syndrome (see 1.3.2), and also cancerous diseases. Activation of AMPK by AICA (AMP analogue) in breast cancer cell lines is able to block proliferation and colony formation in culture, as well as reducing tumor growth in nude mice (Steinberg and Kemp, 2009b).

In *Drosophila,* each subunit of AMPK is encoded by a single gene (Spasić *et al.*, 2009) but alternative transcripts exist and for the γ subunit these encode six different protein isoforms (www.flybase.org). The *loe* mutation affects only one of these isoforms, which is expressed in the nervous system and cannot be replaced by other isoforms (Tschäpe *et al.*, 2002b). Contrary to the fly, in humans the combination of various isoforms for the different subunits can be produced from different genes or by alternative splicing (Hardie, 2007).

4.1 *loe* interferes with the Rho pathway

Experiments done by Tschäpe et al. confirmed that the inhibitory function of AMPK on HMG-CoA reductase is conserved in flies and that alterations in the activity of HMGR, which in *Drosophila* is encoded by the *columbus* gene (Van Doren et al., 1998), play a role in the degenerative phenotype (Tschäpe *et al.*, 2002). As described above, flies do not synthesis cholesterol and therefore the focus of this work is on isoprenoid synthesis, which is another pathway regulated by HMG-CoA. Due to the role of AMPK as a negative regulator of protein prenylation, mutations in this complex are expected to increase isoprenylation and indeed a strong increase in prenylated Rho1 was found, although the membrane-associated, and therefore activated Rho1 only slightly increased (Figure 11 and Figure 12). The small increase in membrane associated Rho1 suggests that isoprenylation of Rho1 might not be the only mechanism that regulates its membrane association. Interestingly, a decrease in the total levels of Rho1 in *loe* mutant flies was observed, indicating that some feed-back mechanism exists to connect the levels of prenylated active Rho1 with Rho1 expression or degradation.

Rho1, a G-Protein and ROCK, the downstream target of Rho1 have essential functions in cell. Rho-GTPases affect many functions of the cellular cytoskeleton depending on the cell type, due to their role in actin polymerization. In neurons, Rho-GTPases have been shown to regulate axon formation and axonal guidance by coupling guidance clues with cytoskeletal rearrangements in the growth cone whereby different family members can differ in their functions (Lowery and Van Vactor, 2009).

Rho1 activation can result in growth cone collapse or promote forward progression depending on its downstream effector (Kranenburg et al., 1999). Rho GTPases are key regulators of the actin and microtubule network and have been connected with several neurodegenerative disease due to increasing data suggesting that impaired axonal transport and neuronal connectivity may be underlying causes of many of these diseases (Chevalier-Larsen and Holzbaur, 2006), (Linseman and Loucks, 2008), (Nadif Kasri and Van Aelst, 2008) (Vickers et al., 2009). By genetic interactions it was shown that *loe* interferes with the Rho1 pathway and its downstream effectors ROCK and LIMK. In agreement with the assumption that *loe* hyperactivates this pathway, mutations in ROCK suppressed the *loe*-induced degeneration (Figure 13). Interestingly ROCK also seems to be correlated with Alzheimer's disease. ROCK has been implicated in the processing of the Amyloid Precursor Protein (APP), from which the neurotoxic Aβ fragments are produced (Cole and Vassar, 2006). Treatment with ROCK inhibitors reduced Aβ levels, whereas increasing prenylation by adding GGPP (geranylgeranyl-pyrophosphate) resulted in increased levels of

Aβ (Zhou et al., 2003).The next target downstream of ROCK is LIMK and as expected neuronal expression of LIMK enhanced the *loe* phenotype (Figure 15). The association of LIMK with a variety of human diseases emphasizes the critical importance of this protein and proper regulation of actin cytoskeletal structures. For example, clusters of p25 (a tubulin polymerization factor that is phosphorylated by LIMK1) are found along the filaments of neurofibrillary tangles in Alzheimer's disease, implicating p25 in tau fibril formation (Lehotzky et al., 2004)

To determine if overexpression of LIMK indeed affects actin polymerization, a globular-/fibrilar- actin assay was used (Figure 18). As expected, the LIMK overexpressing flies have an increase in fibrilar actin, which demonstrates the actin polymerization enhancing function of LIMK. Similarly, fibrilar actin was increased in *loe*, confirming that the *loe* mutation did have the same effect as genetically increasing LIMK.

The *loe* mutation also interacts genetically with cofilin. When twinstar, which encodes cofilin in flies, expression is up- or down- regulated in the *loe* fly, the vacuolization is suppressed in both cases. Because *tsr* flies should reduce levels of active cofilin and TSR overexpression should increase active cofilin, it was expected that the phenotype of *loe* would have been suppressed or enhanced respectively.

This surprising result shows that an alteration of cofilin expression in either direction is beneficial for the neurodegenerative phenotype. Assuming that the activity of cofilin is related to the vacuolization, three possibilities how cofilin could affect the neurodegenerative phenotype in *loe* are possible:

1- Not enough active cofilin:

The upregulation of LIMK in *loe* leads to an increase of phosphorylation of cofilin, which therefore could result in lower levels of active cofilin. Accordingly, an overexpression of constitutively active cofilin improves the phenotype. But a mutation in twinstar, which should lead to an even lower level of active cofilin, does not make it worse.

2- Too much phospho-cofilin:

The other option is that the increased concentration of phospho-cofilin, which will be discussed in 4.2 is causing the phenotype. The *tsr* mutant in the *loe* background could then reduce vacuolization because of a reduction of phospho-cofilin. But if that is the case, the TSR overexpression should have led to an enhancement of the vacuolization due to the consequently elevated p-cofilin levels.

3- Imbalance of active and inactive cofilin :

The results suggest the third theory is the most likely one. That means that the overexpression of cofilin and constitutively active cofilin could lead to an increased active- and phospo- cofilin, therefore reducing the imbalance and improve the phenotype. In this case, overexpressing active cofilin would work even more against the elevated level of phospho- cofilin (as shown in Figure 16– UAS cTSR suppresses the phenotype more than UAS TSR). Generally reducing cofilin by the *tsr* mutant might counterbalance the effect by only effecting phospho-cofilin. Unfortunately there is no antibody against total cofilin available and therefore no testing of this hypothesis possible.

4.2 *loe* has more inactive cofilin

The motility of growth cones is dependent on rapid reorganization of the actin cytoskeleton. When reorganization is not working properly, axonal and dendritic outgrowth is affected.

Surprisingly, removing one copy of cofilin (*tsr*), suppressed the phenotype of *loe* although an enhancement was expected. This was based on the assumption that decreased levels of cofilin expression would aggravate the effects caused by increased LIMK activity and the resulting phosphorylation and inactivation of cofilin by removing even more inactive cofilin. The *loe* fly with the *tsr* mutation also significantly improves the *loe* performance index (Figure 23), which suggests that the tsr mutation is a real suppressor of the *loe* phenotype. Western Blots revealed that the *loe* mutant has 20% more phosphorylated cofilin protein (Figure 17), which confirms the theory that because of hyperactive RHO, *loe* has more active LIMK and therefore more phospho- cofilin. Cofilin gets inactivated by LIMK, which is the contrary reaction to what slingshot does, because slingshot activates cofilin by dephosphorylation. Loe flies also showed a 15% increase in slingshot, indicating that because of the increased level of active LIMK, the *loe* mutant might require a higher amount of slingshot; so more cofilin can be dephosphorylated and activated.

4.3 Cholesterol effects the neurodegeneration phenotype

The brain is the organ with the highest cholesterol level (Dietschy and Turley, 2001). Therefore changes in cholesterol ester (the storage form of cholesterol), which is the case in *loe*, could have a dramatic effect on the maintenance of neurons. As mentioned before, the fruit fly lacks several enzymes required for the *de novo* synthesis of cholesterol. So to maintain the necessary cholesterol levels flies need to obtain it by ingestion. *Loe* has a significant reduction in the cholesterol ester levels, but not free cholesterol levels in the head (Tschäpe et al., 2002b). Therefore it seemed to be interesting to test whether or not a cholesterol free diet would affect the neurodegenerative phenotype in *loe*. Surprisingly, diet did have an effect on the phenotype, but as Figure 30 shows, the cholesterol-free diet causes a significant reduction of vacuolization in male and female flies. This suggests that there is a connection between the cholesterol/ cholesterol ester level and the neurodegeneration. More experiments need to be done to fully understand this correlation.

Although it has been shown that the free cholesterol levels in the *loe* head are unchanged, it is unknown how the concentrations of cholesterol ester and cholesterol are in the fly body. There is a possibility that the *loe* mutation influences cholesterol distribution and that the cholesterol level is elevated in the body and affect behavior. As was shown using rats an overload of free cholesterol in smooth muscle cells (SMC) increases the number of autophagy vesicles (AV), which prevents cell death of SMC (Xu et al., 2010). The protective role on AV of SMC death shows the importance of properly functioning autophagy. Lippai et al. demonstrated that a SNF4A γ *Drosophila* mutant cannot form AV's in the fat body in 3rd larval stage (Lippai et al., 2008).

So it is imaginable that an increased concentration of cholesterol requires an appropriate autophagy response to down regulated the levels, which might not take place in the *loe* fly. That means, that the necessary AV's cannot be formed and reduce the increased cholesterol levels. Therefore, smooth muscle cell death could occur and lead to weak performance in the phototaxis assay (Figure 22).

As shown in Figure 30 the neurodegenerative phenotype improved in *loe* flies that grew up on cholesterol free food. This special diet should have prevented any cholesterol accumulation in the cells, including muscle cells. This suggests, that the improvement should also be found in the fast phototaxis assay, but this still needs to be tested. The lack of a cholesterol could therefore prevent SMC death. However, this needs further investigation.
In addition, hormone sensitive lipase and cholesterol esterase (lipase A) that catalyze the step from cholesterol to cholesterol ester and back, should be tested to understand the connection between cholesterol homeostasis and cell death.

4.4 Neurite outgrowth and transport is affected in *loe*

Neurons have a high energy demand, but a low capacity to store and generate energy, rendering them especially sensitive to changes in energy levels (Poels et al., 2009). Indeed the energy restoring function of AMPK has been shown to play a role in neuronal survival, because a knock-out of the β-subunit (encoded by the *alicorn* gene) in the *Drosophila* eye resulted in light-dependent photoreceptor degeneration (Spasic et al., 2008). A neuroprotective function of AMPK has also been suggested by studies in vertebrate cell cultures.

The experiments involving 3rd instar larva neurons helped to address two issues. First, it showed the transport speed of mitochondria through axons and dendrites of larval neurons and second, an insight of neurite outgrowth in *loe* has been provided. Also the length of the processes is changed compared to wild type. Surprisingly, after 24 hours the *loe* neurons are on average longer. But the growth cannot be maintained, which leads to a reduction of growth in *loe* neurons after 48 hours. Meber and Bamburg (2000) provided direct evidence that increased actin depolymerizing factor/cofilin (ADF) activity by overexpressing ADF/cofilin promotes process extension and neurite outgrowth when measuring the neurite lengths (Meberg and Bamburg, 2000). As previously demonstrated, the *loe* mutant has decreased actin depolymerization, which would be the opposite case of the scenario described by Meber and Bamburg. Therefore this could result in the opposite phenotype, namely reduced outgrowth at least over time. Nevertheless this confirms that altering actin dynamics correlates with changes in neurite outgrowth in *loe*.

Additionally, it is important to remember that the *loe* mutation affects the AMPK γ subunit- where ATP/AMP binding takes place. That means that the energy regulation of the cell is not properly functioning anymore. Therefore a lack of control of ATP consumption is the case. Since anabolic pathways could be without inhibition, a faster outgrowth of processes can be the result (after 24 hours). And the reduced length of processes, after 48 hours, could also be explained with not efficient control of ATP consumption. It seems like after 2 days, most of the ATP got consumed, which leads to a reduced growth.

Tracing mitochondria movement in larval brain neurons by using mitotracker revealed reduced speed of mitochondria moving through *loe* neurites. The velocity was reduced by 42%, which clearly shows that the transport is dramatically disturbed (Figure 19).

Mitochondria, also called "gatekeepers of life and death", are ubiquitous and dynamic organelles involved in many crucial cellular processes in the eukaryotic organism (Correia *et al.*, 2012). As metabolically active cells, neurons have a high-energy demand, which makes them very dependent on functional mitochondria. If mitochondria show a disturbance it can quickly affect the neuronal cell. Neurons are very vulnerable to bioenergetics crisis when it comes to dysfunction of the mitochondrial machinery (Moreira *et al.*, 2009).

It is reasonable to assume that a mutation in the AMPK protein affects mitochondria, considering the fact that AMPK plays a major role in controlling energy productions/consumption within the cell and that mitochondria are responsible for over 90% of cellular ATP production. Therefore, a reduction of the mitochondria velocity through *loe* neurons could be a consequence due to changes in energy metabolism when AMPK is mutated. In addition although *loe* neurites are growing faster in the first 24 hours, which was a really surprising result, the reduction at 48 hours suggests that their structure is compromised presumably also affecting the microtubule network. It is a possibility that the effect on microtubules is indirect and caused by the alteration in actin, because it has been shown that actin guides microtubule into newly forming neurites

(Rodriguez et al., 2003) (Zhou and Cohan, 2004). However it could also be a direct effect.

Nevertheless, it seems unlikely that the formation and extension of microtubules is dramatically affected because the processes are quite capable of growing and it has been shown that growth cones cannot advance without functional microtubules (Tanaka and Kirschner, 1995) (Lowery and Van Vactor, 2009). One could therefore assume that *loe* interferes with the stability of microtubules, which is in agreement with the strong enhancement of the neurodegenerative phenotype of *loe* by a mutation in MAP1b (*futsch^{olk1}*)(unpublished data). MAP1b is a protein known, besides tau, for its microtubule-stabilizing function (Takemura et al., 1992) (Bondallaz et al., 2006). It is possible that AMPK acts directly on MAP1b, because AMPK is closely related to Par-1/MARK (polarity-inducing kinase/microtubule affinity-regulating kinase), which has been shown to phosphorylate MAP's and increases microtubule dynamics (Drewes et al., 1997) (Marx et al., 2010) and affect axonal transport (Mandelkow et al., 2004).

Another explanation for the effects of AMPK on microtubule could be that the effects are mediated through the interference of *loe* with the Rho pathway because it has been shown that LIMK activity can promote microtubule disassembly in endothelial cells (Acevedo et al., 2007), showing a surprising connection between the actin and tubulin cytoskeleton.

Although future experiments are needed to investigate the mechanisms of how AMPK affects microtubule, the results shown in this dissertation demonstrate that interference with isoprenylation and the resulting disruption of the cytoskeleton is an important factor in the progressive neurodegeneration in *loe*.

5 REFERENCES

Acevedo K, Li R, Soo P, Suryadinata R, Sarcevic B, Valova VA, Graham ME, Robinson PJ, Bernard O (2007) The phosphorylation of p25/TPPP by LIM kinase 1 inhibits its ability to assemble microtubules. *Exp Cell Res*, **313:** 4091–4106

Alessi DR, Sakamoto K, Bayascas JR (2006) LKB1-dependent signaling pathways. *Annu Rev Biochem*, **75:** 137–163

Arad M, Benson DW, Perez-Atayde AR, McKenna WJ, Sparks EA, Kanter RJ, McGarry K, Seidman JG, Seidman CE (2002) Constitutively active AMP kinase mutations cause glycogen storage disease mimicking hypertrophic cardiomyopathy. *J Clin Invest*, **109:** 357–362

Bates GP (2005) History of genetic disease: the molecular genetics of Huntington disease - a history. *Nat Rev Genet*, **6:** 766–773

Benzer S (1967) BEHAVIORAL MUTANTS OF Drosophila ISOLATED BY COUNTERCURRENT DISTRIBUTION. *Proc Natl Acad Sci U S A*, **58:** 1112–1119

Bettencourt da Cruz A, Wentzell J, Kretzschmar D (2008) Swiss Cheese, a protein involved in progressive neurodegeneration, acts as a noncanonical regulatory subunit for PKA-C3. *Journal of Neuroscience*, **28:** 10885–10892

Bland ML, Lee RJ, Magallanes JM, Foskett JK, Birnbaum MJ (2010) AMPK supports growth in Drosophila by regulating muscle activity and nutrient uptake in the gut. *Dev Biol*, **344:** 293–303

Blossey R, Schiessel H (2011) The dynamics of the nucleosome: thermal effects, external forces and ATP. *FEBS J*, **278:** 3619–3632

Bondallaz P, Barbier A, Soehrman S, Grenningloh G, Riederer BM (2006) The control of microtubule stability in vitro and in transfected cells by MAP1B and SCG10. *Cell Motil Cytoskeleton*, **63:** 681–695

Brand AH, Perrimon N (1993) Targeted gene expression as a means of altering cell fates and generating dominant phenotypes. *Development*, **118:** 401–415

Brookmeyer R, Johnson E, Ziegler-Graham K, Arrighi HM (2007) Forecasting the global burden of Alzheimer's disease. *Alzheimers Dement*, **3:** 186–191

Burwinkel B, Scott JW, Bührer C, van Landeghem FK, Cox GF, Wilson CJ, Grahame Hardie D, Kilimann MW (2005) Fatal congenital heart glycogenosis caused by a recurrent activating R531Q mutation in the gamma 2-subunit of AMP-activated protein kinase (PRKAG2), not by phosphorylase kinase deficiency. *Am J Hum Genet*, **76:** 1034–1049

Cantó C, Auwerx J (2010) AMP-activated protein kinase and its downstream transcriptional pathways. *Cell Mol Life Sci*, **67:** 3407–3423

Chevalier-Larsen E, Holzbaur EL (2006) Axonal transport and neurodegenerative disease. *Biochim Biophys Acta*, **1762:** 1094–1108

CLARK AJ, BLOCK K (1959) The absence of sterol synthesis in insects. *J Biol Chem*, **234**: 2578–2582

Cole SL, Vassar R (2006) Isoprenoids and Alzheimer's disease: a complex relationship. *Neurobiol Dis*, **22**: 209–222

Correia SC, Santos RX, Perry G, Zhu X, Moreira PI, Smith MA (2012) Mitochondrial importance in Alzheimer's, Huntington's and Parkinson's diseases. *Adv Exp Med Biol*, **724**: 205–221

Crowther DC, Kinghorn KJ, Miranda E, Page R, Curry JA, Duthie FA, Gubb DC, Lomas DA (2005) Intraneuronal Abeta, non-amyloid aggregates and neurodegeneration in a Drosophila model of Alzheimer's disease. *Neuroscience*, **132**: 123–135

Crute BE, Seefeld K, Gamble J, Kemp BE, Witters LA (1998) Functional domains of the alpha1 catalytic subunit of the AMP-activated protein kinase. *J Biol Chem*, **273**: 35347–35354

Culmsee C, Monnig J, Kemp BE, Mattson MP (2001) AMP-activated protein kinase is highly expressed in neurons in the developing rat brain and promotes neuronal survival following glucose deprivation. *J Mol Neurosci*, **17**: 45–58

Deák P, Omar MM, Saunders RD, Pál M, Komonyi O, Szidonya J, Maróy P, Zhang Y, Ashburner M, Benos P, Savakis C, Siden-Kiamos I, Louis C, Bolshakov VN, Kafatos FC, Madueno E, Modolell J, Glover DM (1997) P-element insertion alleles of essential genes on the third chromosome of Drosophila melanogaster: correlation of physical and cytogenetic maps in chromosomal region 86E-87F. *Genetics*, **147**: 1697–1722

Dietschy JM, Turley SD (2001) Cholesterol metabolism in the brain. *Curr Opin Lipidol*, **12**: 105–112

Drewes G, Ebneth A, Preuss U, Mandelkow EM, Mandelkow E (1997) MARK, a novel family of protein kinases that phosphorylate microtubule-associated proteins and trigger microtubule disruption. *Cell*, **89**: 297–308

Fortini ME, Bonini NM (2000) Modeling human neurodegenerative diseases in Drosophila: on a wing and a prayer. *Trends Genet*, **16**: 161–167

Gertler FB, Chiu CY, Richter-Mann L, Chin DJ (1988) Developmental and metabolic regulation of the Drosophila melanogaster 3-hydroxy-3-methylglutaryl coenzyme A reductase. *Mol Cell Biol*, **8**: 2713–2721

Glomset JA, Gelb MH, Farnsworth CC (1990) Prenyl proteins in eukaryotic cells: a new type of membrane anchor. *Trends Biochem Sci*, **15**: 139–142

Goldstein JL, Brown MS (1990) Regulation of the mevalonate pathway. *Nature*, **343**: 425–430

Hardie DG (2007) AMP-activated/SNF1 protein kinases: conserved guardians of cellular energy. *Nat Rev Mol Cell Biol*, **8**: 774–785

Hawley SA, Davison M, Woods A, Davies SP, Beri RK, Carling D, Hardie DG (1996) Characterization of the AMP-activated protein kinase kinase from rat liver and identification of threonine 172 as the major site at which it phosphorylates AMP-activated protein kinase. *J Biol Chem*, **271**: 27879–27887

Hudson ER, Pan DA, James J, Lucocq JM, Hawley SA, Green KA, Baba O, Terashima T, Hardie DG (2003) A novel domain in AMP-activated protein kinase causes glycogen storage bodies similar to those seen in hereditary cardiac arrhythmias. *Curr Biol*, **13**: 861–866

Inglese J, Koch WJ, Caron MG, Lefkowitz RJ (1992) Isoprenylation in regulation of signal transduction by G-protein-coupled receptor kinases. *Nature*, **359**: 147–150

Ji H, Ramsey MR, Hayes DN, Fan C, McNamara K, Kozlowski P, Torrice C, Wu MC, Shimamura T, Perera SA, Liang MC, Cai D, Naumov GN, Bao L, Contreras CM, Li D, Chen L, Krishnamurthy J, Koivunen J, Chirieac LR, Padera RF, Bronson RT, Lindeman NI, Christiani DC, Lin X, Shapiro GI, Jänne PA, Johnson BE, Meyerson M, Kwiatkowski DJ, Castrillon DH, Bardeesy N, Sharpless NE, Wong KK (2007) LKB1 modulates lung cancer differentiation and metastasis. *Nature*, **448**: 807–810

Jovceva E, Larsen MR, Waterfield MD, Baum B, Timms JF (2007) Dynamic cofilin phosphorylation in the control of lamellipodial actin homeostasis. *J Cell Sci*, **120**: 1888–1897

Kranenburg O, Poland M, van Horck FP, Drechsel D, Hall A, Moolenaar WH (1999) Activation of RhoA by lysophosphatidic acid and Galpha12/13 subunits in neuronal cells: induction of neurite retraction. *Mol Biol Cell*, **10**: 1851–1857

Lehotzky A, Tirián L, Tökési N, Lénárt P, Szabó B, Kovács J, Ovádi J (2004) Dynamic targeting of microtubules by TPPP/p25 affects cell survival. *J Cell Sci*, **117**: 6249–6259

Linseman DA, Loucks FA (2008) Diverse roles of Rho family GTPases in neuronal development, survival, and death. *Front Biosci*, **13**: 657–676

Lippai M, Csikós G, Maróy P, Lukácsovich T, Juhász G, Sass M (2008) SNF4Agamma, the Drosophila AMPK gamma subunit is required for regulation of developmental and stress-induced autophagy. *Autophagy*, **4**: 476–486

Lowery LA, Van Vactor D (2009) The trip of the tip: understanding the growth cone machinery. *Nat Rev Mol Cell Biol*, **10**: 332–343

Luo L (2000) Rho GTPases in neuronal morphogenesis. *Nat Rev Neurosci*, **1**: 173–180

Mahlapuu M, Johansson C, Lindgren K, Hjälm G, Barnes BR, Krook A, Zierath JR, Andersson L, Marklund S (2004) Expression profiling of the gamma-subunit isoforms of AMP-activated protein kinase suggests a major role for gamma3 in white skeletal muscle. *Am J Physiol Endocrinol Metab*, **286**: E194–200

Marx A, Nugoor C, Panneerselvam S, Mandelkow E (2010) Structure and function of polarity-inducing kinase family MARK/Par-1 within the branch of AMPK/Snf1-related kinases. *FASEB J*, **24**: 1637–1648

Meberg PJ, Bamburg JR (2000) Increase in neurite outgrowth mediated by overexpression of actin depolymerizing factor. *J Neurosci*, **20**: 2459–2469

Meyer-Lindenberg A, Mervis CB, Berman KF (2006) Neural mechanisms in Williams syndrome: a unique window to genetic influences on cognition and behaviour. *Nat Rev Neurosci*, **7**: 380–393

Moreira PI, Duarte AI, Santos MS, Rego AC, Oliveira CR (2009) An integrative view of the role of oxidative stress, mitochondria and insulin in Alzheimer's disease. *J Alzheimers Dis*, **16**: 741–761

Nadif Kasri N, Van Aelst L (2008) Rho-linked genes and neurological disorders. *Pflugers Arch*, **455**: 787–797

Ng J, Luo L (2004) Rho GTPases regulate axon growth through convergent and divergent signaling pathways. *Neuron*, **44**: 779–793

Niwa R, Nagata-Ohashi K, Takeichi M, Mizuno K, Uemura T (2002) Control of actin reorganization by Slingshot, a family of phosphatases that dephosphorylate ADF/cofilin. *Cell*, **108**: 233–246

Novelli G, D'Apice MR (2012) Protein farnesylation and disease. *Journal of inherited metabolic disease*,

Pandey UB, Nichols CD (2011) Human disease models in Drosophila melanogaster and the role of the fly in therapeutic drug discovery. *Pharmacol Rev*, **63**: 411–436

Querfurth HW, LaFerla FM (2010) Alzheimer's disease. *N Engl J Med*, **362**: 329–344

Reiter LT, Potocki L, Chien S, Gribskov M, Bier E (2001) A systematic analysis of human disease-associated gene sequences in Drosophila melanogaster. *Genome Res*, **11**: 1114–1125

Rodriguez OC, Schaefer AW, Mandato CA, Forscher P, Bement WM, Waterman-Storer CM (2003) Conserved microtubule-actin interactions in cell movement and morphogenesis. *Nat Cell Biol*, **5**: 599–609

Scott RW, Olson MF (2007) LIM kinases: function, regulation and association with human disease. *J Mol Med (Berl)*, **85**: 555–568

Sidhu JS, Rajawat YS, Rami TG, Gollob MH, Wang Z, Yuan R, Marian AJ, DeMayo FJ, Weilbacher D, Taffet GE, Davies JK, Carling D, Khoury DS, Roberts R (2005) Transgenic mouse model of ventricular preexcitation and atrioventricular reentrant tachycardia induced by an AMP-activated protein kinase loss-of-function mutation responsible for Wolff-Parkinson-White syndrome. *Circulation*, **111**: 21–29

Spasić MR, Callaerts P, Norga KK (2009) AMP-activated protein kinase (AMPK) molecular crossroad for metabolic control and survival of neurons. *Neuroscientist*, **15**: 309–316

Steinberg G, Kemp B (2009a) AMPK in Health and Disease. *Physiol Rev*, **89**: 1025–1078

Steinberg G, Kemp B (2009b) AMPK in Health and Disease. *Physiol Rev*, **89**: 1025–1078

Strauss R, Heisenberg M (1993) A higher control center of locomotor behavior in the Drosophila brain. *J Neurosci*, **13**: 1852–1861

Suter M, Riek U, Tuerk R, Schlattner U, Wallimann T, Neumann D (2006) Dissecting the role of 5'-AMP for allosteric stimulation, activation, and deactivation of AMP-activated protein kinase. *J Biol Chem*, **281**: 32207–32216

Takemura R, Okabe S, Umeyama T, Kanai Y, Cowan NJ, Hirokawa N (1992) Increased microtubule stability and alpha tubulin acetylation in cells transfected with microtubule-associated proteins MAP1B, MAP2 or tau. *J Cell Sci*, **103**: 953–964

Tanaka E, Kirschner MW (1995) The role of microtubules in growth cone turning at substrate boundaries. *J Cell Biol*, **128**: 127–137

Tschäpe JA, Hammerschmied C, Mühlig-Versen M, Athenstaedt K, Daum G, Kretzschmar D (2002a) The neurodegeneration mutant löchrig interferes with cholesterol homeostasis and Appl processing. *EMBO J*, **21**: 6367–6376

Tschäpe JA, Hammerschmied C, Mühlig-Versen M, Athenstaedt K, Daum G, Kretzschmar D (2002b) The neurodegeneration mutant löchrig interferes with cholesterol homeostasis and Appl processing. *EMBO J*, **21**: 6367–6376

Vickers JC, King AE, Woodhouse A, Kirkcaldie MT, Staal JA, McCormack GH, Blizzard CA, Musgrove RE, Mitew S, Liu Y, Chuckowree JA, Bibari O, Dickson TC (2009) Axonopathy and cytoskeletal disruption in degenerative diseases of the central nervous system. *Brain Res Bull*, **80**: 217–223

Wolozin B, Kellman W, Ruosseau P, Celesia GG, Siegel G (2000) Decreased prevalence of Alzheimer disease associated with 3-hydroxy-3-methyglutaryl coenzyme A reductase inhibitors. *Arch Neurol*, **57**: 1439–1443

Xu K, Yang Y, Yan M, Zhan J, Fu X, Zheng X (2010) Autophagy plays a protective role in free cholesterol overload-induced death of smooth muscle cells. *J Lipid Res*, **51**: 2581–2590

Zhou FQ, Cohan CS (2004) How actin filaments and microtubules steer growth cones to their targets. *J Neurobiol*, **58**: 84–91

Zhou Y, Su Y, Li B, Liu F, Ryder JW, Wu X, Gonzalez-DeWhitt PA, Gelfanova V, Hale JE, May PC, Paul SM, Ni B (2003) Nonsteroidal anti-inflammatory drugs can lower amyloidogenic Abeta42 by inhibiting Rho. *Science*, **302**: 1215–1217

6 ABBREVIATIONS

Abbreviation	
AD	Alzheimer's disease
ADF	actin depolymerizing factor
AICA	5-aminoimidazole-4-carboxamide
AMP	adenosine monophosphate
AMPK	AMP- activated protein kinase
Appl	Amyloid precursor protein like
ATP	Adenosine triphosphate
CaMKK	Calmodulin-dependent kinase kinase
CBS	cystathionine-b-synthase
CSP	85ysteine string protein
DNA	Deoxyribonucleic acid
F-actin	filamentous actin
FPPS	Farnesyl- Pyrophosphat-Synthase
G-actin	globular actin
GAP	GTPase-activating protein
GDP	guanosine diphosphate
GEF	Guanine nucleotide exchange factor
GLUT4	glucose transporter 4
GTP	guanosine triphosphate
HMGR	HMG-CoA-Reductase
LIMK	LIM Kinase
LKB1	liver kinase B1
MARK	microtubule affinity-regulating kinase
MAP	Microtubule-associated protein
NTE	Neuropathy target esterase
PD	Parkinson's disease
PDVF	Polyvinylidene fluoride
PJS	Peutz–Jeghers syndrome

Abbreviations

PP	pyrophosphate
ROK	Rho kinase
SDS	sodium dodecyl sulfate
SNF1	sucrose non-fermenting 1
SSH	Slingshot
TAK1	TGF-b-activated kinase-1
TSR	twin star
UAS	Upstream Activating Sequence
WPW	Wolff-Parkinson-White syndrome
20 HE	20-hydroxyecdysone

LIST OF PUBLICATIONS

Publications relevant to this thesis

Cook M, Mani P, Wentzell JS, Kretzschmar D (2012) The neurodegenerative *loechrig* (loe) mutant interferes with RhoA prenylation and the integrity of the cytoskeleton - **accepted- PLOS ONE**

Abstract released for poster presentations

- 2011- **Mandy Cook**, Jill Wentzell, Doris Kretzschmar. The neurodegenerative *loe* mutant interferes with the RHO Pathway. *"Drosophila Research Conference-2011"*- San Diego- USA

- 2011- **Mandy Cook**, Jill Wentzell, Doris Kretzschmar. The neurodegenerative *Drosophila Melanogaster* mutant *loe* interferes with the RHO Pathway.
"Student Research Forum" -Oregon Health and Science University- Portland- USA

- 2012- **Mandy Cook**, Jill Wentzell, Doris Kretzschmar. The neurodegenerative AMPK mutant *loe* interferes with the RHO Pathway and actin dynamics.
"Drosophila Research Conference-2012"- Chicago- USA

- 2012- **Mandy Cook**, Jill Wentzell, Doris Kretzschmar. The neurodegenerative AMPK mutant *loe* interferes with the RHO Pathway and actin dynamics.
  "*Innovation and Opportunities in Rare Disease Research*" Oregon Health and Science University- Portland- USA
- 2012- **Mandy Cook**, Doris Kretzschmar. The neurodegenerative AMPK mutant *loe* interferes with the RHO Pathway and actin dynamics.
  "*2012 Neurofly meeting* "- Padua- Italy

Abstract released for oral presentations

- 2012- **Mandy Cook**, Doris Kretzschmar. The neurodegenerative AMPK mutant *loe* interferes with the RHO Pathway and actin dynamics.
  "*OHSU–Research Week*"- Oregon Health and Science University- Portland- USA

MoreBooks! publishing

i want morebooks!

Buy your books fast and straightforward online - at one of world's fastest growing online book stores! Environmentally sound due to Print-on-Demand technologies.

Buy your books online at
www.get-morebooks.com

Kaufen Sie Ihre Bücher schnell und unkompliziert online – auf einer der am schnellsten wachsenden Buchhandelsplattformen weltweit! Dank Print-On-Demand umwelt- und ressourcenschonend produziert.

Bücher schneller online kaufen
www.morebooks.de

VDM Verlagsservicegesellschaft mbH
Heinrich-Böcking-Str. 6-8   Telefon: +49 681 3720 174   info@vdm-vsg.de
D - 66121 Saarbrücken   Telefax: +49 681 3720 1749 www.vdm-vsg.de

Printed by Books on Demand GmbH, Norderstedt / Germany